# GEOLOGY, ENVIRONMENT AND SOCIETY

## ABOUT THE AUTHOR

K.S. Valdiya is Indian National Science Academy Golden Jubilee Research Professor at the Jawaharlal Nehru Centre for Advanced Scientific Research, Bangalore. He has over forty-five years of research and teaching experience, and has been the recipient of a number of awards—including the S.S. Bhatnagar Prize from the Council of Scientific and Industrial Research, the L. Rama Rao Gold Medal from the Geological Society of India, the Pitambar Pant National Environment Fellowship from the Ministry of Environment and Forests, the Wadia Medal from the Indian National Science Academy, the National Mineral Award of Excellence from the Ministry of Mines, Government of India, and the S.K. Mitra Award of the Indian National Science Academy. Prof. Valdiya is a Fellow of the Third World Academy of Sciences, Indian National Science Academy, Indian Academy of Sciences and National Academy of Sciences as well as Honorary Fellow of the Geological Society of America and the Nepal Geological Society.

Prof. Valdiya has authored eight books and monographs, edited nine books and written over 90 research papers and 40 popular articles.

# GEOLOGY, ENVIRONMENT AND SOCIETY

K.S. Valdiya

Universities Press

**Geology, Environment and Society**

UNIVERSITIES PRESS (INDIA) PRIVATE LIMITED

*Registered Office*
3-6-747/1/A & 3-6-754/1, Himayatnagar, Hyderabad 500 029 (A.P.), India
e-mail: info@universitiespress.com

*Distributed by*
**Orient Blackswan Private Limited**

*Registered Office*
3-6-752, Himayatnagar, Hyderabad 500 029 (A.P.), India

*Other Offices*
Bangalore / Bhopal / Bhubaneshwar / Chennai
Ernakulam / Guwahati / Hyderabad / Jaipur / Kolkata
Lucknow / Mumbai / New Delhi / Patna



First published 2004
Reprinted 2011

ISBN: 978 81 7371 505 1

*Typeset by*
Trinity Designers & Typesetters
Chennai 600 041

*Printed in India at*
Graphica Printers
Hyderabad 500 013

*Published by*
Universities Press (India) Private Limited
3-6-747/1/A & 3-6-754/1, Himayatnagar
Hyderabad 500 029 (A.P), India

For Indira

To whom, very simply, I owe everything

# Acknowledgements

I am deeply grateful to the President and the administration of the Jawaharlal Nehru Centre for Advanced Scientific Research, for providing extremely generous support, and extending all facilities to help me write this book on environmental geology.

Grateful thanks are due to Dr. Dalim K. Paul (Kolkata), Prof. H.Y. Mohan Ram (New Delhi), Dr. L.M.S. Palni (Kosi-Katarmal), Dr. M. Ramakrishna (Bangalore), Prof. S.K. Tandon (New Delhi), Prof. S.L. Kayastha (Varanasi), Prof. V. Subramanian (New Delhi), Dr. C.D. Thatte (New Delhi), Prof. S.C. Sarkar (Kolkata), Prof. O.P. Varma (Roorkee), Dr. J.R. Kayal (Kolkata), Dr. C.P. Rajendran (Thiruvananthapuram), Dr. R.C. Lakhera (Dehradun), Dr. M.P. Sah (Dehradun), Dr. R.K. Bhandari (Chennai), Prof. Vishwas S. Kale (Pune), Prof. Rajiv Sinha (Kanpur), Dr. O.N. Dhar (Pune), Prof. U.C. Mohanty (New Delhi), Major General S.S. Sharma (Chandigarh), Dr. A.K. Tangri (Lucknow), Dr. R.N. Singh (Nagpur), Prof. B.C. Raymahashay (Kanpur), Prof. R.N. Iyengar (Bangalore), Dr. R. Anbalagan (Roorkee), Dr. P.C. Nawani (Dehradun), Shri Gyaneshwar Pant (Lucknow), Shri Rajendra Sanwal (Lucknow), Prof. K.C. Sahu (Mumbai) and Dr. A.V. Shekdar (Nagpur) for their thorough review of relevant chapters and for giving invaluable suggestions for improvement. I am deeply indebted to these friends.

I am very thankful to Prof. R.P. Gupta (Roorkee), Prof. L.S. Chamyal (Vadodara), Dr. C.P. Rajendran (Thiruvananthapuram), Dr. A.K. Tangri (Lucknow), Dr. J.T. Gergan (Dehradun), Major General S.S. Sharma (Chandigarh), Dr. D.P. Rao (Hyderabad), Dr. A.S. Rajawat (Ahmadabad), Prof. M.J. Haigh (Oxford), Dr. G.C.S. Negi (Shrinagar) and Dr. S.K. Bartarya (Dehradun) for the photographs that illustrate the book.

10 May 2004

K.S. Valdiya

# Contents

# Preface

At the stroke of midnight on 15 August 1947, when India woke to freedom (borrowing Pandit Jawaharlal Nehru's words), about 360 million people inhabited the newly independent, vast country. More than half a century later, with India poised on the threshold of a technological revolution, the population has grown to one billion, of which over 320 million are so poor that they can hardly afford the basic nourishment of two square meals a day. For these acutely impoverished people the main concern is survival in the present—the fragility of the country's resource-base, the diminishing beauty of its landscape, or the increasing impurity of its rivers are not likely to be their immediate concerns. This is the tragedy of an explosive population growth, and the resultant increase in the number of the underprivileged.

To compound this bitter tragedy, people at the other end of the spectrum—the powerful corporate sector—exploit and appropriate the country's natural resources and cover the land with ugly and dangerous wastes. People in authority look away unconcerned, while environmental calamities spiral and play havoc with the plight of the people.

The consequences are for all to see. We are selling our future to finance our present—plundering resources that must be conserved, dirtying waters that are already scarce, fouling air that is thick with pollutants, and defacing a landscape that was once exquisitely beautiful. India's tryst with freedom was meant to make the country safe, secure and progressive. What we have actually done, and continue to do today, is to make our present convenient, while making our future insecure.

There is indeed an all-round *growth*, but no *development*. Even if there has been development, it has failed to reduce or eliminate poverty. 'It [has] only modernized it,' as Ivan Illich puts it so aptly. Despite a considerable wealth of natural resources in the country and a rich human talent, many people are deprived and live in the fear of ecological disasters. A majority of Indians now wait to face the consequences of environmental stresses that their own omissions and commissions have engendered. As misfortune will have it, those who conceive, plan and implement development programmes do not discern the virtues in conserving the gifts of nature, preserving the health of ecosystems, exercising caution while acting in geodynamically sensitive areas, making efforts to restore damaged environment, or making provisions to share the benefits of development amongst all sections of the society, including development oustees, eco-refugees and the socially deprived.

It is the tyranny of small decisions that lets one small section of the society reap all the benefits, while the majority bears the consequences of environmental deterioration and its consequent hazards. It is nothing but tyranny that, while the entire burden of environmental and socio-cultural changes resulting from the construction of a dam, is borne by people in the river catchment, the entire benefit goes to the people living downstream in the command area. It is doubtless a small decision to allow valuable forests to be destroyed to feed paper mills; or to let roads be constructed indiscriminately and unscientifically on unstable hillslopes in order to open an area for marketing. Industries and mines have been dumping their wastes, without the slightest effort to utilize usable submarginal grades, and without taking measures to check pollution of riverwater. Decisions are often taken to meet political expediency, or to realize immediate objectives, without taking into consideration long-term implications. This often results in squalour in the midst of edifying magnificence.

Efforts made to tame rivers in flood, to check erosion along storm-struck coasts, to mitigate hazards of slope failures, etc, have not always met with success. Not only has the labour been wasted, the money invested has also turned infructuous. This is because there is little attempt to understand the natural causes of the geological hazards or to understand the varying cycles and processes of nature under stress. Moreover, there is scant appreciation and consideration for communities that will have to withstand environmental stresses and natural hazards in their daily lives. Our action plans entail merely 'responses to crises', rather than measures to make the communities resilient to disasters. In other words, we have ignored the time-tested premise of not interfering with protective environmental systems, and also ignored the conservation of natural features—dunes that protect the coasts from sea erosion, wetlands in alluvial plains that accommodate floodwater, forest cover that absorbs excess rainwater and also prevents the erosion of topsoil in hillslopes, and so on.

This book is a modest attempt at addressing environmental problems that are intimately related to the framework of geology. It deals with issues that concern environmental security in the context of geological settings and geodynamic sensitivities of the terrain. The book also offers strategies and measures to cope with the consequences of tampering with the geological balance.

This work represents the perceptions of a foot-soldier among geologists, who has endeavoured to apply the principles and practices of geology for the improvement and preservation of the environment, through the judicious use of natural resources, and by ensuring that safeguards exist against degradation of land, contamination of air, and pollution of water.

One of the concerns of geology, to my mind, is to protect society from the risks of natural hazards. This endeavour embraces the whole gamut of living off the earth—recognizing, assessing and coping with natural hazards, and applying careful technology and geotechnics so that damage to the environment and harm to the affected communities is reduced to a minimum. Being well-equipped to assess the infirmities and capabilities of the land, geologists are better placed to guide the process of selection of appropriate sites for buildings, bridges and dams, and for choosing suitable sites for waste disposal.

I believe that a geologist can play a crucial role in ensuring a harmonious relationship between human beings and their natural environment, and in assisting in the application of appropriate technologies for the benefit of humans and their surroundings, both animate and inanimate.

10 May 2004 K.S. Valdiya

Mourn not the dead,
But rather mourn the apathetic throng—
The cowed and meek,
Who see the world's great anguish and its wrong
And dare not speak.

— Ralph Chaplin

# 1

# Human Beings: An Ecological Perspective

## Humans and the Environment: Scale of Interaction

Human beings may consider themselves the most competent and resourceful of all creatures on earth, but they are still helplessly dependent on other organisms for their existence. This is because humans, unlike plants, cannot produce their own food, and have to depend on plants and other animals for their subsistence.

Plants, microbes and animals constitute living or **biotic** communities. Biotic communities are intimately associated and inextricably linked with non-living or **abiotic** communities like soil, water and air. These living and non-living components function inter-dependently as ecological systems or **ecosystems**. Energy—flowing from a variety of sources, including the sun—and various biochemical processes enable the ecosystems to function. Biochemical processes cycle nutrients, involving carbon, oxygen, nitrogen, phosphorus and sulphur, besides minerals like iron, copper, manganese, boron, etc. Naturally, ecosystems would not function if the energy flow were to slow down or stop. And, if ecosystems were to malfunction, the integrity of all biotic communities would be in peril.

It is evident that human beings, the environment (with its web of ecosystems) and the earth system[1] are inseparably and inter-dependently linked. If one is in distress, the other two will also be affected seriously (Figure 1.1). If human beings wish to live a life free of suffering, they will have to ensure the integrity of ecosystems, to make sure that the land is not damaged, the water and air are not polluted. The *Yajurveda* articulates an ideal attitude towards the environment and its living components—

दृते दृह्न मा मित्रस्य मा चक्षुषा
सर्वाणि भूतनि समीक्षन्ताम्
मित्रस्याऽञ्ह चक्षुषा सर्वाणि भूतानि समीक्षे
मित्रस्य चक्षुषा समीक्षामहे ।

May all beings look on me as a friend.
May I look on all beings as friends.
May we all look on one another with friendship.

— *Yajurveda* (36:18)

[1] The earth system includes the lithosphere (the earth's crust and upper mantle), hydrosphere (groundwater, surfacewater, oceanwater), biosphere (animal and plant worlds) and atmosphere.

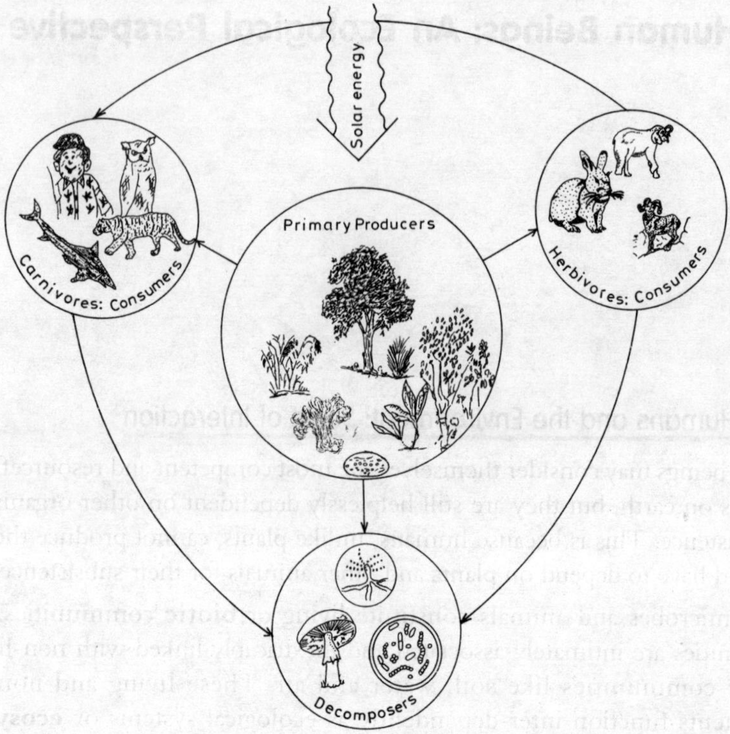


**Figure 1.1.** Humans, the environment and the earth system are inextricably linked and function inter-dependently. Living communities comprise producers (plants, cyanobacteria, etc), consumers (fungi, non-autotrophic bacteria, animals, etc) and decomposers (bacteria, fungi, etc).

## Vulnerable Humans

Human beings not only have the ability to adapt themselves to varying environmental conditions, but also to manipulate ecosystems to their benefit. Consequently, they occupy a pre-eminent position among living communities.

Living communities comprise three groups of organisms (Figure 1.1). Plants and autotrophic bacteria are self-nourishing organisms. They use sunlight, water and simple inorganic substances to synthesize complex food substances such as carbohydrates, proteins and lipids. They are therefore called **producers**. Being incapable of manufacturing their food, animals, including humans, consume plants and other animals. They are the **consumers**. Fungi and other microbes are called **decomposers** as they feed on living or dead organisms (Figure 1.1). Producers, consumers and decomposers constitute the web of living ecosystems.

Humans are tied to other living beings through the **food chain** (Figure 1.2). The food chain shows humans to be utterly dependent on plants and other animals for their food, and therefore for their survival. It follows that the well-being of humans depends inexorably on the stability of other biotic communities—on the integrity of ecosystems made up of plants, animals and microbes.

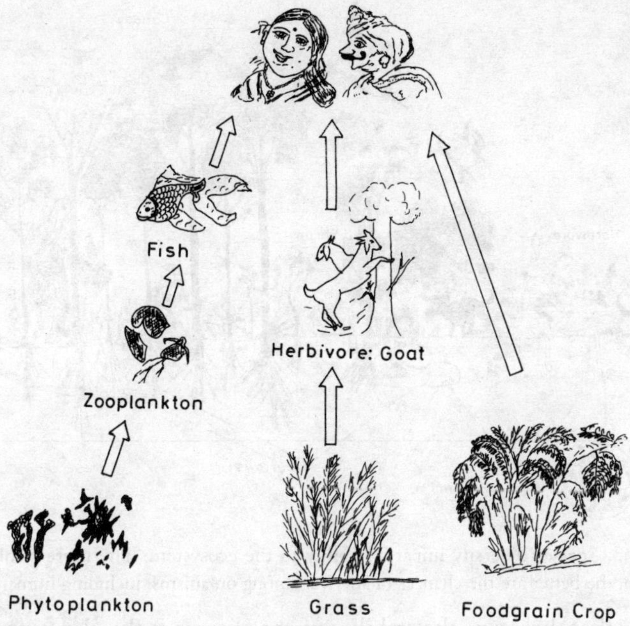


**Figure 1.2.** Even though they occupy the top position in the food chain, humans are utterly dependent on plants, microbes and other animals for food, and therefore for survival.

## Stability of Ecosystems

Diversity of species and genera among communities of organisms imparts strength and stability to ecosystems. The greater the diversity, the greater the ability of the community to adapt itself to the stress of change, and hence, the stabler the ecosystem. Large numbers of species or genera provide insurance against decline and extinction (Figure 1.3). The evolutionary process, resulting from the mixing and rearrangement of genetic material, depends on the availability of a wide genetic base. Furthermore, some species do manage to survive disease, poisoning, drought, fire, physical assault, or serious distress. Those that survive, pass on their characteristics of tolerance and resistance to their descendants. That is why there is a strong need to preserve **genetic diversity** in both natural and human-made ecosystems—islands, biosphere reserves, wildlife sanctuaries, agricultural plantations and fields, parks, rivers, lakes, etc. Any action or change that leads to a decrease in species diversity—like the monoculture of select varieties of rice, wheat, etc—would lead to the decline and eventual demise of the ecosystem.

Ecosystems with a limited number of species, such as monocultural cultivations of pine and eucalyptus, are not only very vulnerable to fire and disease, but are also poor in promoting groundwater recharge. The dry, needle-shaped pine leaves form an impermeable layer when they fall down, and do not permit rainwater to percolate into the ground. Furthermore, the oil-rich leaves do not decompose, to create the humus which promotes water percolation. Pine forests are therefore devoid of shrubs, undergrowth or grass cover and there is very little groundwater underneath.

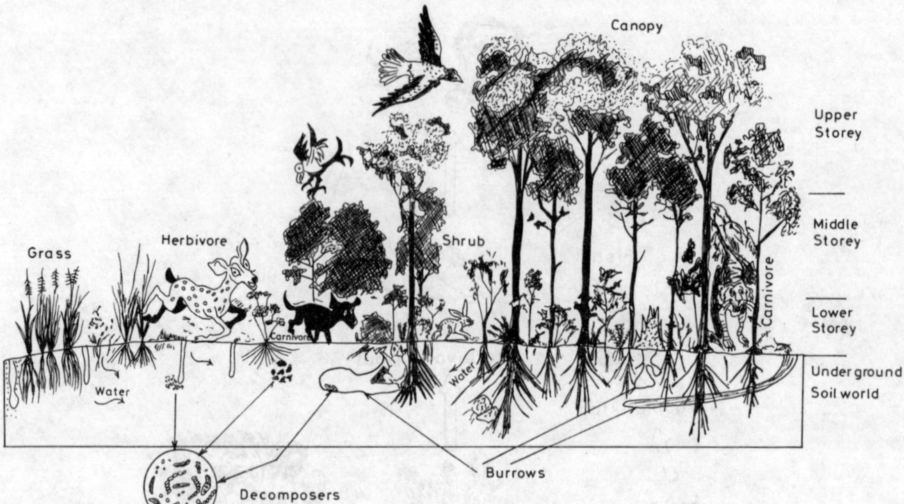


**Figure 1.3.** Species diversity imparts stability to the ecosystem. The more stable the ecosystem, the better are the chances of survival among organisms, including humans.

Another factor that threatens the stability of ecosystems is the alarming increase in human population. In India, this trend is disturbing (Figure 1.4). More than one billion people inhabit a land that is already under severe stress. Nearly 320 million people—36 per cent of India's total population—live below the poverty line, and an equal number hover just above that line. The limits of the land's carrying capacity has been reached, and these ecosystems cannot withstand any further exploitation and assault.

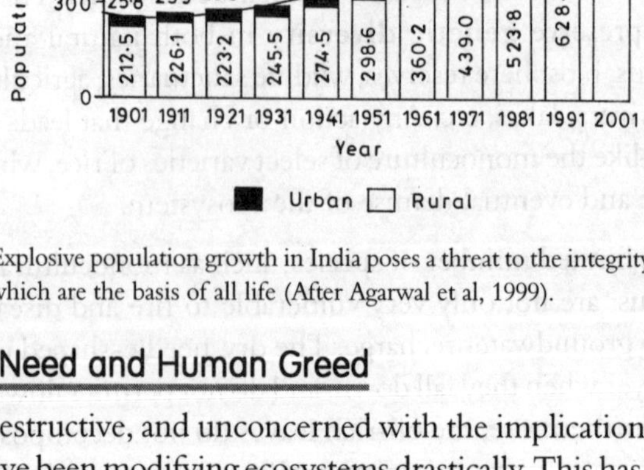


**Figure 1.4.** Explosive population growth in India poses a threat to the integrity and safety of ecosystems which are the basis of all life (After Agarwal et al, 1999).

## Human Need and Human Greed

Avaricious and destructive, and unconcerned with the implications of their actions, human beings have been modifying ecosystems drastically. This has led to unforeseen and often unintended changes, with far-reaching consequences. Orissa can be used as a case in example. The large-scale killing of riverine crocodiles resulted in an economic crisis that had not been foreseen. The disappearance of the crocodiles led to the disappearance of fish, which in turn had dire implications for people dependent on fishing for their livelihood. It was later learnt that the crocodiles used to stir up sediments and keep the water free from clogging, thus enabling aquatic plants to grow. These aquatic plants had supported large populations of fish and other organisms.

The ecosystem comes under assault from three sections. The underprivileged living on the fringes of forests are inclined to scrounge the forests for firewood, fodder and other necessities, resulting in the overuse, and therefore abuse of fragile ecosystems. The powerful and the rich manipulate the environment in a manner that leaves resources depleted—the landscape defaced, the water poisoned and the air polluted. Planners, who are supposedly well-equipped to carry out development work, often disregard environmental security, with little understanding of the dynamics of earth systems. They may often create adverse affects through the implementation of well-meaning programmes. **Sustainable development** is the only answer to check human beings who destroy their tomorrows, in order to live today.

# 2

# Geology: An Agenda for Earth Care, Environmental Security and Safe Society

## Understanding Geology

The nature and functions of various ecosystems can be understood from the geodynamic processes at work in the earth. The professional application of geology ensures a sustainable endeavour for development—a development which does not necessitate adverse effects on the earth's resource base, or injury to the ecosystems. Indeed, all geological pursuits should be seen as exercises towards locating natural resources for the care of the earth and its various systems (Figure 2.1).

The paramount objective of geological pursuits is to find clues that help locate natural resources, and maintain the earth as a safe habitat for human beings and their fellow organisms. Geologists, of necessity, study changes that are taking place in the earth and endeavour to understand the implications of these, as well as past occurrences, in order to foretell what could happen in future. The principal aim of these exercises is to provide guidelines for human activities that will not upset the environmental balance.

## Crucial Items on the Agenda

One of the most precious assets on earth is land (Plate 1). Making use of this asset in accordance with its capabilities is an important assignment for geologists. **Land capability** depends on the carrying capacity of the ecosystem. It is necessary to know the nature of the soil that supports vegetation and permits construction of various utility structures on it.

Equally important is knowledge of the condition of rocks—particularly their strength to withstand the loads and associated stresses of structures like high-rise buildings, dams, bridges, roads, canals, airfields, etc. Geotechnical investigations based on geological, geophysical and geomorphological studies are pre-requisites for all development activities that are likely to have an impact on the integrity and stability of the terrain and its ecosystems. Engineering geology or **geotechnics** (Figure 2.1) is the forte of earth science—it is the essence of all geological pursuits that result in better quality of life in societies.

## Hazard Mitigation Exercises

The mounting pressures of fast-growing populations have forced people to occupy and develop areas prone to natural hazards. For the safety of the community, it is necessary to identify hazard-prone areas and assess the risks to life and property (Figure 2.1). This can be done by perceiving the threats involved when making artificial changes in the terrain, and also predicting the probable behaviour of hazardous events. Assessing and estimating acceptable risks govern the formulation of **Disaster Preparedness Plans** and the measures to be taken for mitigating damaging effects.

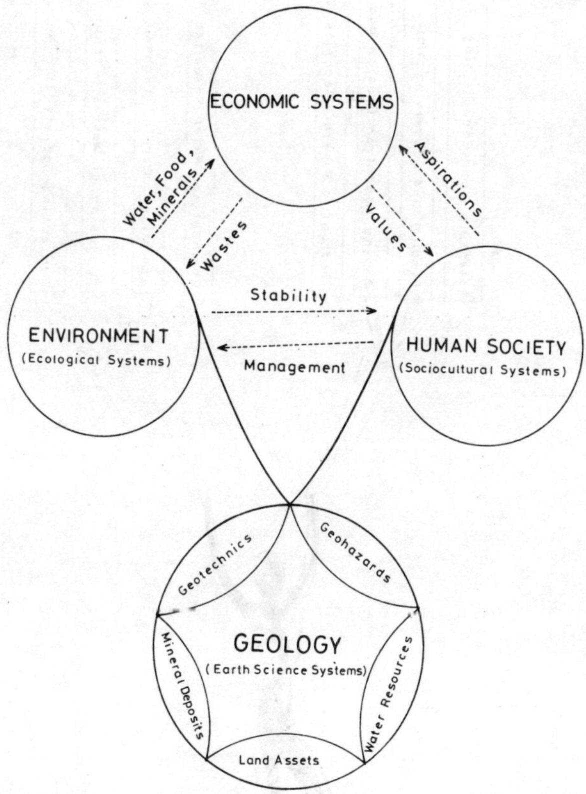


**Figure 2.1.** Dimensions of geology—the facets of geological studies that are relevant to the well-being, safety and security of society.

## Targeting and Conserving the Earth's Resources

The supply of life-supporting water, in adequate quantity and of requisite quality, is essential for living communities (Figure 2.1). Water should be obtained using scientifically appropriate methods, from surface and underground sources, and managed properly. The geological problems related to the distribution of underground water, flow of surface water, excessive use of water in floodplains and the crises caused by declining discharges in hill springs have highlighted the crucial importance of **hydrogeological** investigations.

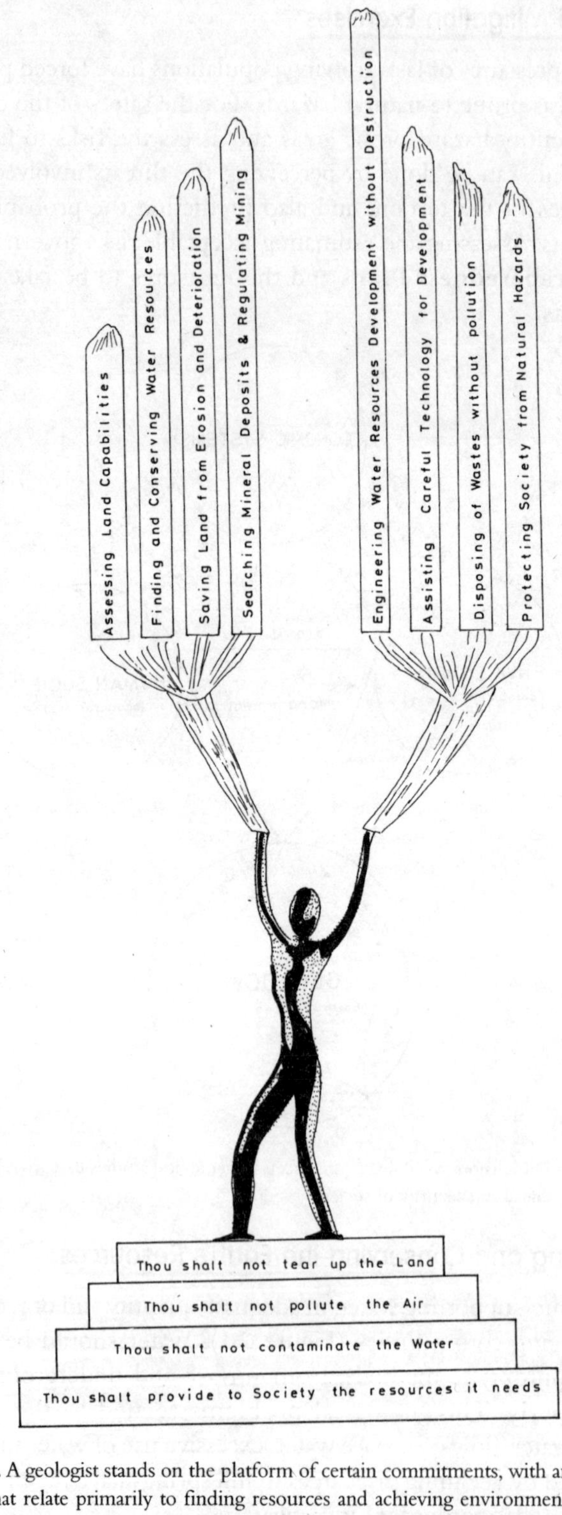


**Figure 2.2.** A geologist stands on the platform of certain commitments, with an agenda of activities that relate primarily to finding resources and achieving environmental security. A true geologist is an earth-caring and community-serving scientist.

Minerals are essential for a large number of industries, which need predictable and plentiful supplies of these 'anchor stones' of development. The management of **mineral resources** (Figure 2.1) entails not only identifying the location and extent of mineral deposits, but also devising methods of mining (extraction) without causing harm to the environment. **Economic geologists** aim to conserve the depleting reserves of minerals (Figure 2.2).

As the conventional sources of energy get depleted, and in some cases practically exhausted, it is important that new sources of minerals and energy are identified, located and their potential estimated. One of these new sources of energy could be hot springs in geodynamically active zones. Likewise, the presence of gas hydrates in the continental slopes off the coast of the Western Ghat, and in the Andaman islands, have spurred fresh geological investigations in the oceanic realm.

Industries and power plants, as also fast-expanding urban centres, need intelligent guidance to select sites for the **disposal of wastes** without poisoning the soil and polluting the water.

## Earth-caring and Community-serving Scientists

Of great and critical import is the problem of how to manage natural resources—soils, water, minerals and forests. The pace and dimension of industrial-economic development requires a national level board for **Integrated Natural Resources Management**—which entails planning and coordinating all efforts towards exploration and the use of forest, water, mineral and soil resources (Narain, 1995). There is not only need for large-scale surveys to locate and expand mineral resources, but also to carry out basin-wise studies on water-budgeting and to develop water-management techniques.

The role of the geologist in society is that of a person who actively and critically participates in all programmes of development. In other words, the true geologist is an earth-caring and community-serving scientist—one who helps to understand, preserve, ameliorate and restore the natural environment, and helps to nurse natural assets stored in rocks and sediments (Figure 2.2).

# 3

# Knowing Land-capability: Caring for Cherished Assets

## A Precious Resource

The traditional Hindu *bhumipooja* (worship of the earth) before the start of construction work is symbolic of an adoration of land, as a cherished asset (Plate 1). In the previous chapter, it was identified that optimal use of land must be made in accordance with its capability and suitability, and must be well within its carrying capacity. The capability of any land depends upon its geological setting under the ground, and its biological condition above the ground. Hill tracts, for example, are more capable of supporting forests, while floodplains should ideally be used for cultivation of crops. Barren lands, on the other hand, can be used to locate urban centres, set up industries and power plants, and if the situation demands, submerged in order to build artificial reservoirs.

Whatever its use, the integrity of this precious and limited resource needs to be preserved at all costs. Indigenous farmers have for long known the ethos of appropriate land use. This is evident from the sustainability of farming through the ages—a result, obviously, of effective conservation measures.

**Table 2.1.** Pattern of land use in India (in million hectares) (Bhan and Singh, 1994).

| | |
|---|---|
| Total geographic area of India | 328.73 m ha |
| Reported area for land utilization | 304.89 m ha |
| Forest land | 115 m ha |
| Permanent pastures and grazing land | 2.2 m ha |
| Agricultural land | 150 m ha |
| Cultural wasteland, barren and uncultivated land | 16.5 m ha |
| Land put to non-agricultural uses | 21.2 m ha |

### TOO MANY PEOPLE, TOO LITTLE LAND

Today, much of the land in India is being exploited far beyond its carrying capacity. Admittedly there are few options, since Indian population is growing at an explosive rate and the availability of land is fast diminishing. The laws of inheritance have led to property being divided into smaller uneconomic holdings. The **per capita land**

**available** was 0.89 ha in 1951, and came down to 0.33 ha in 2000. Likewise, the per capita cultivable land-holding dwindled from 0.48 ha in 1951 to 0.15 ha in 2000. The per capita pastoral land for cattle decreased from 0.37 ha in 1951 to just 0.10 ha in 2000 (Bhan and Singh, 1994).

It also appears that the **tragedy of common** has overwhelmed Indian society. Deep in people's minds is a feeling that land belonging to the public does not deserve the care of society—a notion that 'property belonging to everybody, is property belonging to nobody'. This ruinous misconception has left many public lands, which should have been the concern of society, derelict.

## Carrying Capacity of Land

Land is the surface expression of its underlying geology. The management of land must therefore be based on understanding its lithological–structural setting, and the geomorphological layout of the area. This means: (i) a thorough assessment of the potential—capability and capacity—of the land to support the planned activity, (ii) demarcation of the identified land, based on the specific purpose it suits, and (iii) comparative analysis of seasonal alternatives, in view of problems likely to arise from the specific land use.

The important actions needed to prepare a **land-use management plan** are:

(i) Inventorying land resources through the interpretation of precise maps (including aerial and satellite imagery), and then delineating the requisite information on large-scale topographic maps.

(ii) Conducting comprehensive investigations to bring out geomorphological, geological, hydrological, pedological and ecological characterizations of the area.

(iii) Identifying areas vulnerable to natural hazards (including abnormal soil behaviour) and spelling out the likelihood of their occurrence, and anticipated intensity and frequency in the times to come.

(iv) Assessing the likely impact that land use will have, based on the monitoring of geomorphological, geophysical, hydrological, pedological and ecological changes occurring in the present time.

## Land-use Planning: Regional and Urban

A developing country requires a comprehensive and sound land-use management programme. This should take into consideration the broader aspects of development—agricultural land, urban centres, industries, water resources development projects (dams and reservoirs), transportation networks (roads and railway lines), airfields and seaports, mining for minerals, forest conservation, defence requirements (cantonments and troop deployment), etc. In other words, regional land-use planning is a pre-requisite for a nation's development planning.

Very soon, nearly half the population of India will be living in cities and towns.

Urban centres need to be well-planned if their inhabitants are to live a decent life. Urban land-use planning encompasses habitat planning on the basis of: (i) condition/stability of the ground in order to construct utility structures, (ii) absence of natural hazards like ground-subsidence, landslides, floods, cyclones, etc, (iii) availability of water of requisite quality and in sufficient quantity, (iv) availability of sites suitable for the disposal of all kinds of wastes, and (v) proximity of sites for recreation, water supply, etc.

With rare exceptions, most Indian cities have become dense conglomerations of buildings and utility structures. Traffic lanes are choked and there is little or no space left for recreational facilities and aesthetic pursuits. There is not enough place for waste disposal. More than 30 per cent of the urban population lives in slums. Moreover, the opportunity to improve and innovate the design of a town or city is limited, or absolutely lacking. Cities are growing, but at the expense of productive agricultural land. In the last fifty years, more than 2.5 million hectares of agricultural fields were lost to urban centres. These centres have an enormous appetite for all kinds of resources, and in turn generate great volumes of garbage and other wastes.

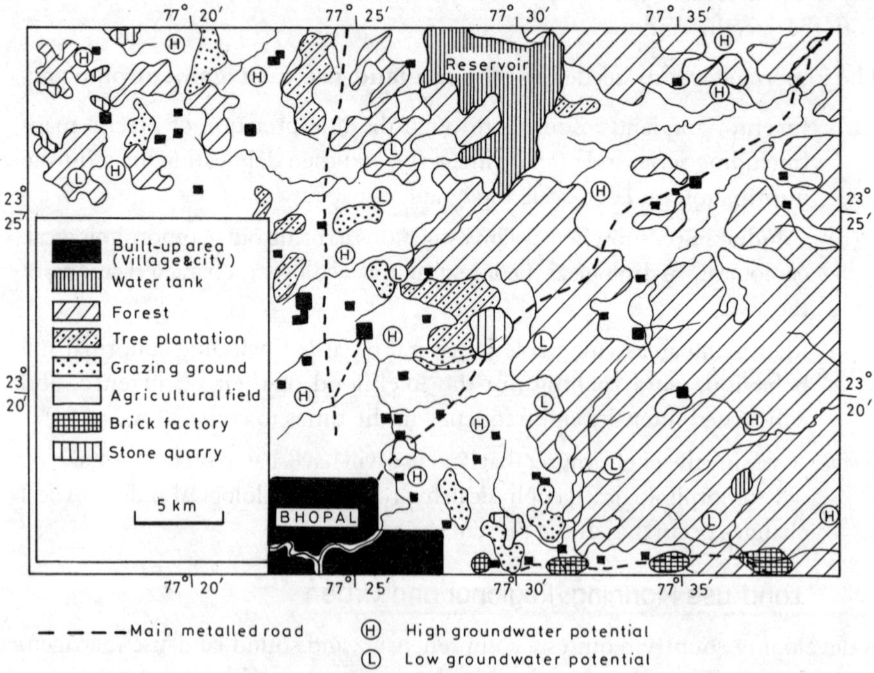


**Figure 3.1.** Land-capability map of an urban area—Bhopal in Madhya Pradesh (Cited in Valdiya, 1987).

## THE NEED FOR A PARADIGM SHIFT IN LAND-USE POLICY

The influx of people from rural regions to urban centres is uncontrollable, and unabated. If this exodus is not reversed, cities could become 'megaslums'. In 2001, the population living in urban slums in India numbered 40.6 million against the total urban population of 285.3 million. People living in such slums constitute 49 per cent (5.8 million) of the total urban population in Mumbai, 19 per cent (1.9 million) in New Delhi, 33 per cent (1.5 million) in Kolkata, and 26 per cent (1.1 million) in Chennai (*The Hindu*, 27 November 2001). There is, therefore, a fervent need for a paradigm shift in **land-use policy**.

To control this, the peripheral areas of the country can be opened up for new settlements, and also for the resettlement of development-refugees. Such settlements can be developed around tree farms, refineries, power plants, fertilizer factories, mineral-processing plants, dams and reservoirs, all of which can provide employment to the displaced people. People living in hazard-prone areas, and those who have been displaced for various reasons (the submergence of land behind dams, for example) can be resettled in new, outlying sites—far from the present urban centres. Ideally, such people should not only be given some land to settle down in, but also provided with loans, and amenities like schools, hospitals, parks, theatres, transportation services, telecommunication facilities, etc. These amenities must be developed by the government or the institutions responsible for uprooting the people.

Also, the facilities existing in rural centres needs to be strengthened and improved, with all the civic amenities that cities command, so that there is only a limited exodus to urban centres.

# 4

# Nursing Soil Sickness and Treating Land Injury

## Soil: A Living Ecosystem

Like a precious carpet spread over the land, the soil sustains life by providing nourishment to plants, storing and providing passage to groundwater, providing shelter to burrowing creatures, and supporting the foundations of buildings and utility structures. Soil comprises sand, silt and clay, and a variety of chemical compounds that act as valuable nutrients—all of which are products formed by the weathering of rocks that lie beneath the soil. These are intermixed with humus and other organic matter, contributed primarily by rotting roots, stems and leaves of plants, and soil fauna (Figure 4.1).

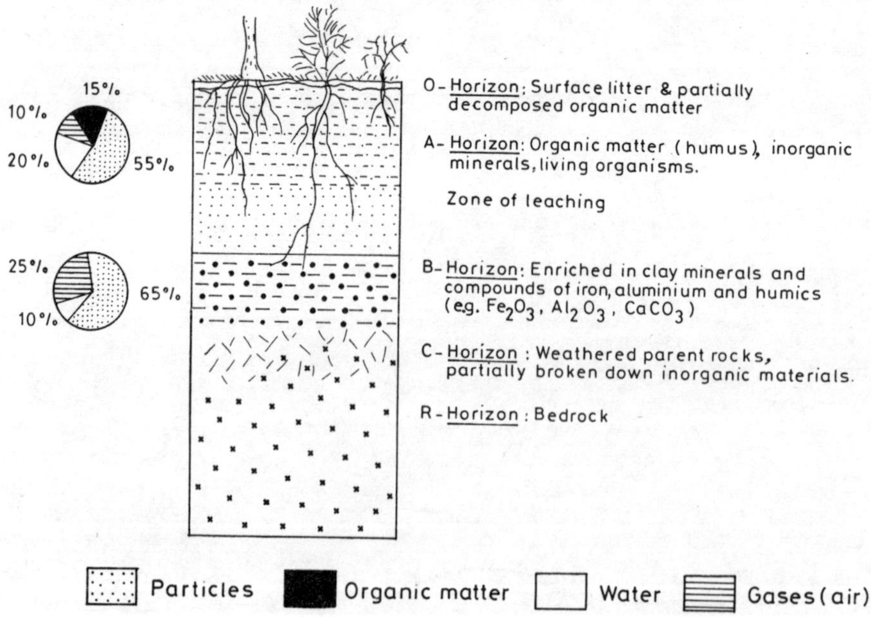


**Figure 4.1.** Integrally connected with the weathered rocks beneath and the vegetation above, the soil is a living ecosystem characterized by its own microclimate. The microclimate is created by the air and water that fills the interstitial space. On the basis of compositional characteristics, four horizons of soil are recognized.

Humus imparts permeability to the soil, allowing water to pass through the interspaces between soil particles. Both humus and clay particles have the capacity to attract water. Molecules of water form a rim of wetness around soil particles, creating an 'atmosphere' of water vapour and air within the soil. This keeps the soil moist even during dry seasons and allows life to establish a foothold and grow. Since air and water fill the interstitial space within, the soil has a microclimate of its own. This microclimate is not only conditioned by water and air, but is also controlled by the organisms that live within, and upon, the soil (Figure 4.2).

Humus provides organic food, while clay and inorganic components derived from rock-weathering provide nutrients to the organisms and plants that depend on the soil. The presence of moisture and nutrients promotes microbial activity within the soil, making the soil a living ecosystem. It is this ecosystem on which rests the foundations of very many other ecosystems.

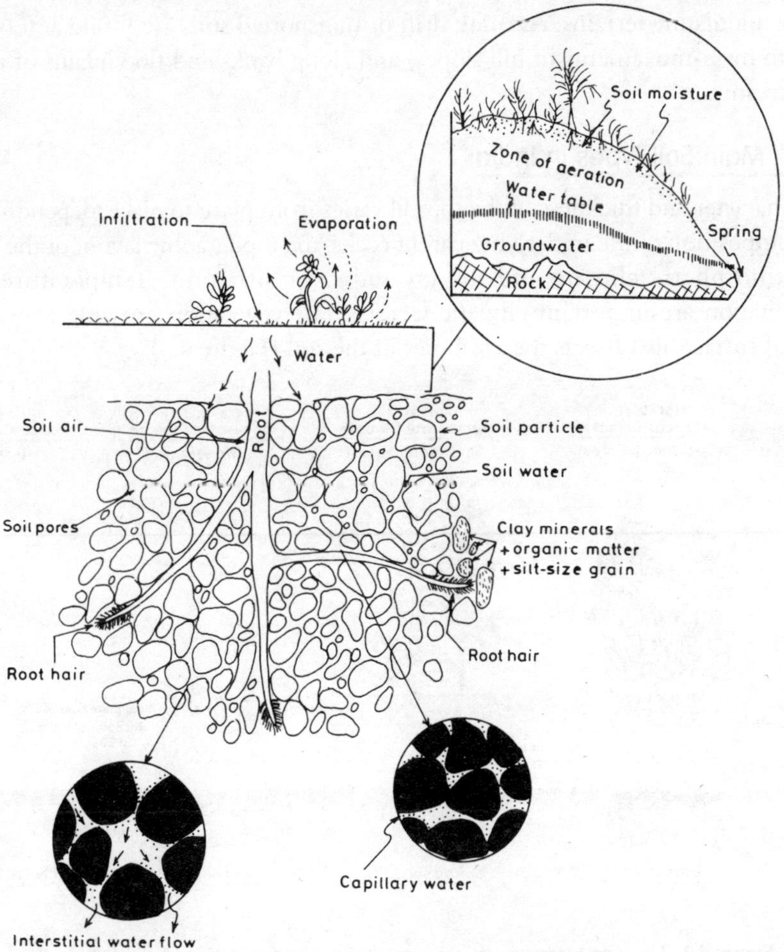


**Figure 4.2.** Presence of water, air and nutrients (derived from rock-weathering), and organic matter (formed primarily by rotting vegetation) promotes microbial activity within the soil and supports plant life above it. *Inset* shows zones of soil filled with air and moisture.

## FORMATION OF SOILS

The weathered rocks grade upwards into the regolith or **C-Horizon** (Figure 4.1), which is made up of sand-sized particles, silt and clay. This gives way further upwards to the **B-Horizon**, characterized by various oxides that accumulate as a result of downward leaching of mineral-rich water. The oxides represent secondary enrichment through percolation of water. The transitionally overlying **A-Horizon** is deficient in leached-out material but rich in humus, moisture and air. It is also known as **topsoil**. The topsoil supports all kinds of life—fungi, bacteria, plants and burrowing animals.

Formation of soil is a slow process. Nature takes more than a thousand years to form a centimetre-thick soil profile.

In terms of occurrence, two types of soils are recognized. **Zonal** or *in situ* soils (formed as a result of weathering of rocks) occur in their place of origin in relatively flat or undulating terrains. **Azonal**, drift or transported soils are found at foothills (due to mass-movements in hill slopes) and along banks and floodplains of rivers and streams.

## Main Soil Types in India

The character and thickness of the topsoil varies from place to place, depending on the composition of the underlying parent rocks, the topographic layout of the land, the atmospheric climate and the groundwater situation. Temperature and precipitation are important climatic factors. The vegetation growing above the ground surface also affects the attributes of the soil (Figure 4.3).

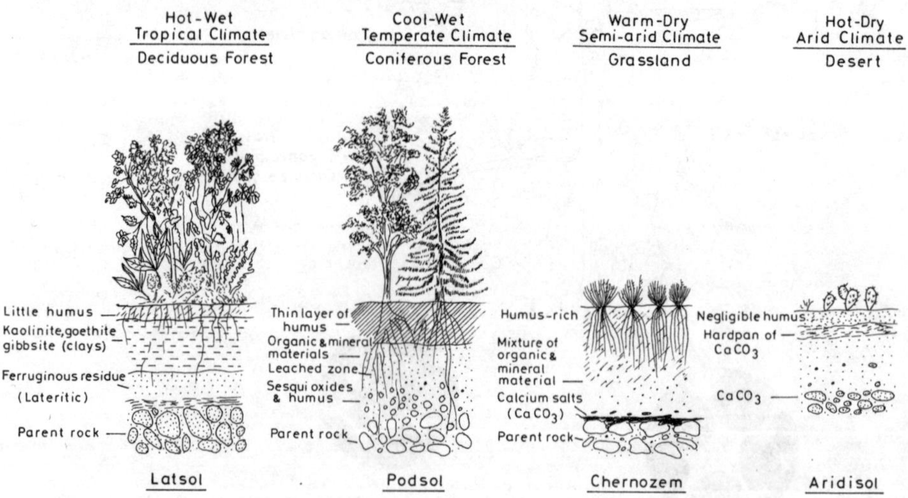


**Figure 4.3.** The type (character) of soil varies from region to region, depending on the composition of the underlying (bed) rocks, topographic setting of the terrain, groundwater situation and climatic conditions of the region. These are the main types of soils that have been recognized in the Indian subcontinent (Based on Valdiya, 1987).

In the hot-humid climate of tropical regions, the organic matter derived from luxuriously growing deciduous forests accumulates and alters very rapidly—developing the soil profile in a span of a few hundred years. The presence of humus not only increases fertility of the soil, but also imparts a degree of permeability to it. This facilitates movement of water and air within the soil. The bacterial activity is very vigorous, resulting in considerable transformation of the organic matter. The soil is consequently deficient in the organic acids that humus would have otherwise generated. Under circumstances of heavy rainfall, the bases are leached out, and sesquioxides of iron and aluminium, together with oxides and hydroxides of iron, become concentrated. The iron oxides occur as pellets or tiny spheres (≡pisolites). The resulting soil is known as **latsol** or **laterosol** (Figure 4.3). Laterosol, which has very low fertility, covers very large parts of eastern, central and southern India (Figure 4.4).

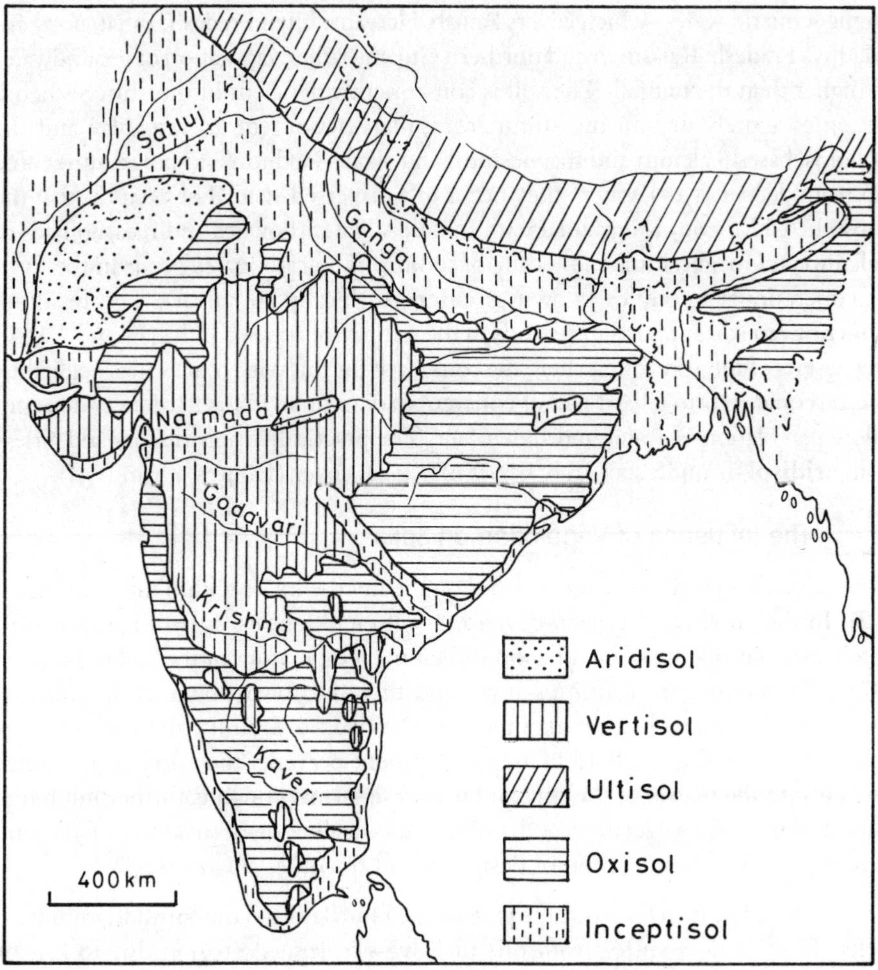


**Figure 4.4.** Distribution of the main types of soils in the Indian subcontinent (Based on a map in the *National Atlas of India*, National Atlas and Thematic Maps Organization, Kolkata).

In the cool-wet climate of the temperate zone, organic acids are produced during humification, which leach out compounds of calcium and magnesium, leaving silica in abundance. The base-deficient, acidic, ash-grey soil is called **podsol**. Podsol has rather low fertility and supports only coniferous forests in large parts of the Lesser Himalaya and elsewhere (Figure 4.4).

*Regur* is a unique type of black, loamy soil that covers practically the whole of the Malwa and Deccan plateaus (Figure 4.4). A **vertisol** of sorts, it is the product of the weathering of basaltic lavas and associated pyroclastic rocks of the Deccan Volcanic Province. Regur consists of 40–48% montmorillonite clay, 6–8% carbonates of calcium and/or magnesium, less than 10% aluminium sesquioxides, 9–10% iron oxides, and less than 0.5% alkalis and phosphates. It is very fertile and supports rich crops of cotton, sugarcane, groundnut, and soyabean, apart from the grasslands and savanna forests of the Deccan.

In the semiarid zone—which covers Panjab, Haryana, Uttar Pradesh, Bihar, northern Madhya Pradesh, Rajasthan and northern Gujarat—the evaporation of groundwater is higher than the rainfall. The soil is consequently deficient in moisture. When it becomes acutely dry during summer, capillary action comes into play and the colloids, bases (calcium and magnesium) and alkalis (sodium and potassium) come up from deeper layers, where they occur in a dispersed state. The result is that the topsoil (A-Horizon) is enriched with minerals like carbonates and bicarbonates of calcium and magnesium, and locally of sodium. The carbonates occur in the form of concretionary nodules known as **calcrete** (locally as *kankar*). The resulting **chernozem** soil supports grasslands in the temperate zone of India (Figure 4.4). In the deserts of the arid zone, where the content of humus within the soil is negligible, the carbonates form a hard pan of concretionary nodules. This hard layer does not allow percolation of water and flow of air. The practically moistureless and airless soil, **aridisol**, is understandably bereft of vegetal cover (Figures 4.3 and 4.4).

## The Influence of Vegetation on Soil

Vegetation also exercises strong control on the microclimate within the soil (Figure 4.2). In the absence of vegetal cover, the soil rapidly heats up and is exposed to unchecked beating by rain and wind. When the temperature of the soil rises above 40°C, the production of humus stops, and there is a breakdown of the nutrient cycle and the microclimatic equilibrium. Deprived of humus, the soil becomes impermeable and quite hard. This development prevents the entry of rainwater and air into the matrix of the soil, and therefore, plant growth is further inhibited. Interdependent as vegetation and soil are, the decline or destruction of the one brings about the degeneration or destruction of the other (Figure 4.5A).

Once covered by lush forests, large parts of the Himalaya and the Sahyadri mountain belts (Plate 2) now suffer from this malady—accelerated erosion due to loss of vegetal cover. Extensive deforestation has diminished and locally destroyed the fertility of these soils, their shear strengths and their ability to soak up rainwater.

The appearance of drought-tolerant, xerophytic plants bears eloquent testimony to the drastic loss of soil moisture in the deforested tracts of these two mountain domains. The plateaus and plains of Peninsular India are equally deprived of the protection of vegetation (Plate 2). This explains the increase in wastelands (14 m ha) and uncultivable barren lands (19 m ha) in the Indian subcontinent.

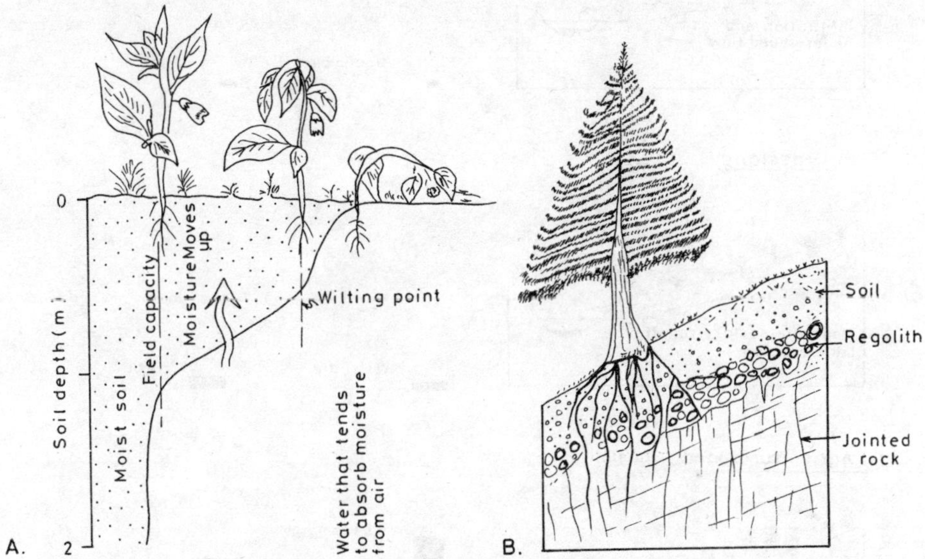


**Figure 4.5.** (A) Rise of temperature beyond 40°C stops production of humus—causing not only deficiency in nutrients, but also decrease in the permeability of the soil. In the absence of water and air within the soil, plants wilt and die.

(B) Penetrating through the soil and thrusting through cracks in the bedrock, roots impart strength to the soil. The loss of vegetation results in weakening of the soils' shear strength, leading to denudation and erosion.

The loss of vegetation has far-reaching consequences. The roots of trees, shrubs and grasses hold the soil together and provide it strength (Figure 4.5B). Decay of roots weakens the shear strength of the soil. This makes the soil vulnerable to erosion and slipping. The hazards of denudation and mass-movements increase considerably when and where the soil is bereft of vegetation. Further, the soil surface becomes prone to the impact of raindrops and the action of water torrents. In the absence of foliage which acts as an umbrella of sorts, there is no interception of rainfall and the ground becomes vulnerable to erosion and denudation. Any action that removes or reduces the cover of vegetation—over-grazing, deforestation, excavation for mining, or road construction—promotes or augments erosion and gully formation. This manifests itself in the generation of larger and larger quantities of detritus carried by streams, and in progressively increasing the amount of rainwater flowing over the ground as runoff—without percolating underground (Figure 4.6).

In a greatly deforested (@ 0.57% per year) watershed in eastern Nepal, the loss of topsoil increased from 12.5 tonnes/ha/yr in 1989 to 18 tonnes/ha/yr in 1991 (Sah et

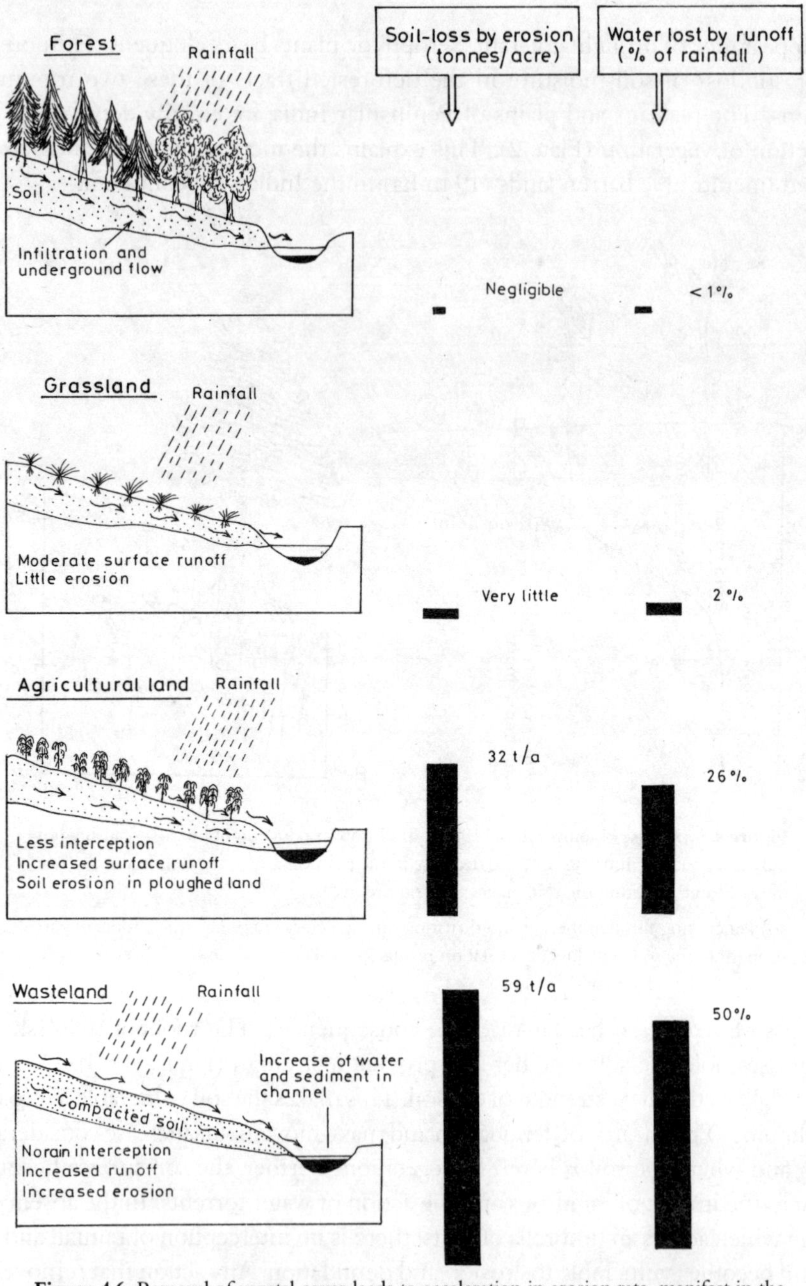


**Figure 4.6.** Removal of vegetal cover leads to acceleration in erosion rate, manifest in the generation of larger and larger quantities of detritus. Rainwater runoff also increases progressively in the absence of underground percolation in vegetation-deprived lands.

al, 1979). Forest clearing in the Cherrapunji hills in Meghalaya—which registers the highest rainfall in the Indian subcontinent—has resulted in losing much of the rainfall as storm runoff and leaving little water in the soil for the dry season. The result is that the springs and wells of the hills have gone dry, and the Bangladesh plains are increasingly plagued by floods during the monsoon.

Unsound agricultural practices—such as upslope extension of farming without terracing, *bunding* or building drainage ditches on unstable slopes in the Himalayan region, and the slash-burn-shift cultivation (*jhum*) practiced in the hills of northeast India—have greatly weakened the stability of hill slopes. This is manifest in enormous loss of soils. A study showed that the soil-loss from the *jhum* land was 7.24 t/ha/yr, in comparison to 0.76 t/ha/yr in the case of grass-covered tracts (Ramakrishnan and Tokay, 1978). In a central Kumaun catchment, while the slopes covered by oak and pine forests were eroded at the rate of 0.02–0.04 mm/yr, the deforested tract and the farmed lands were denuded at a rate four times higher: 0.09–0.18 mm/yr (Rawat and Rawat, 1994). In Sikkim the conversion of dense mixed forest into open degraded forest increased soil-loss from 4.18 t/ha/yr in 1988 to 8.82 t/ha/yr in 1992 (Rai and Sharma, 1998). Along with the erosion of soil, the nutrient loss amounted to 305 t/ha/yr.

Raindrops create millions of tiny craters on bare ground bereft of vegetation. The craters connect to form rills. Torrents of rainwater turn the rills into gullies, and the gullies eventually become ravines. This process of progressive accentuation of gully action leads to the development of ravine land, also called **badland**, (Figure 4.7 and Plate 3) as seen in the valleys of the rivers Mahi in Gujarat, Chambal in Madhya Pradesh and Damodar in West Bengal.

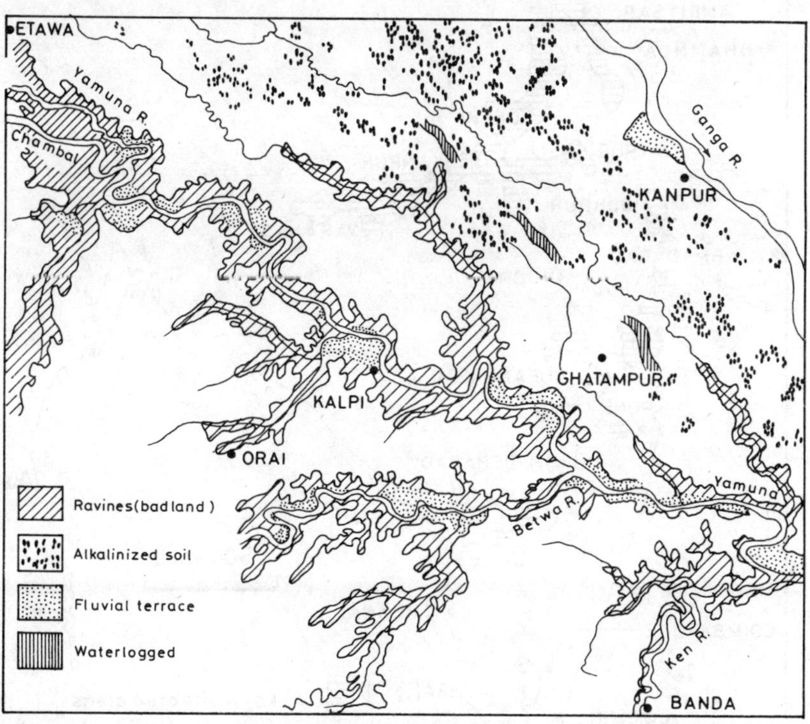


**Figure 4.7.** The Yamuna and its tributaries, including the Betwa and the Ken in Uttar Pradesh and adjoining Madhya Pradesh, are characterized by spectacular development of badlands. (After Figure 1 in page 191 in *Contributions in Environmental Geology*, Geological Survey of India, Kolkata, 1996)

Deprived of foliage cover, the dry lands of Rajasthan and northern Gujarat are ravaged by wind-induced deflation of the loose dry soil. The wind sweeps away finer particles and deposits them in the form of dunes in faraway places, while the coarser material is left behind as skeletal drift heaps. Nearly 38.7 m ha of Indian land is afflicted by wind erosion (Abrol, 1994).

## Soil Sickness

As already stated, in arid and semiarid tracts where evaporation exceeds rainfall, capillary action brings up colloidal material and salts to the soil surface from deeper layers within the soil. This includes predominant carbonates and bicarbonates of calcium. The carbonates are deposited in the form of concretionary nodules called calcrete or *kankar*. In some places, carbonates and bicarbonates of sodium ($Na_2CO_3$, $NaHCO_3$), and smaller quantities of sulphates and chlorides of sodium, magnesium and calcium are brought up. This has given rise to very serious soil diseases like **alkalinization** and **salinization**. As little as 0.2% of sodium carbonate is enough to make the soil quite infertile. Obviously, alkalinization and salinization have rendered unproductive very large tracts of agricultural lands (Figure 4.8). Known in different places as *usar, kallar, chopan, kshar, chandu, thur, reh* and *kar*, the sick soil

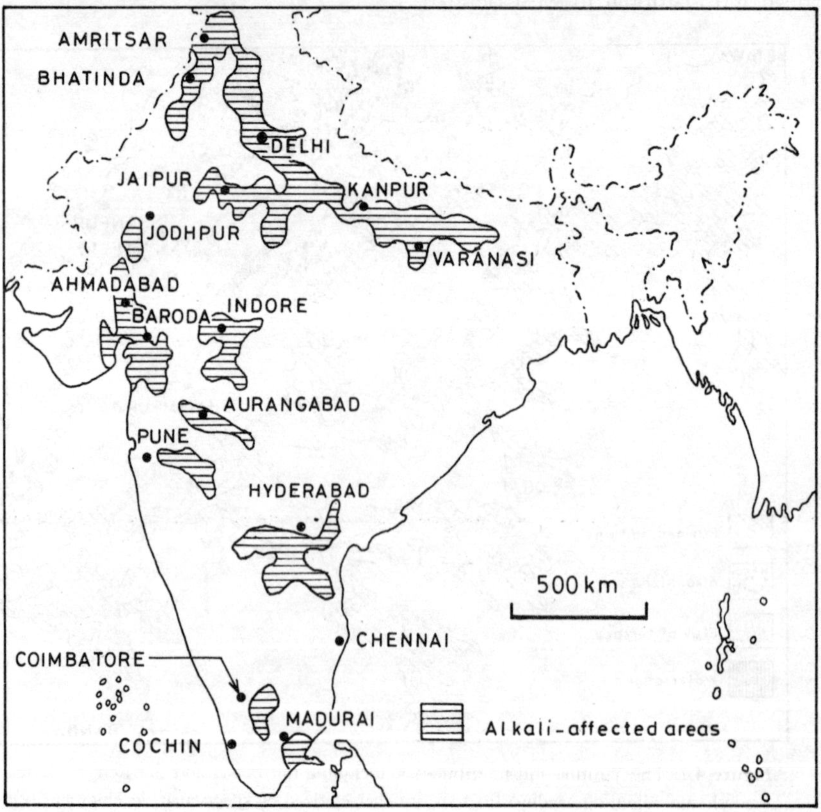


**Figure 4.8.** Alkali-affected, over-irrigated, but poorly drained areas in India (Based on Mehta, 1983).

covers more than 11 m ha of the Indian landmass—which means that more than 8% of the geographical area of India suffers from the sickness of salinization and alkalinization (Table 4.1).

**Table 4.1.** Extent of salt- and alkali-affected areas in India (in ha) (After G. Singh, 1994).

| State/ Union Territory | Alkali-affected area | Alkali- + Salt-affected area | Salt-affected area |
|---|---|---|---|
| Panjab | 495,000 | | 25,000 |
| Haryana | 275,000 | | 185,000 |
| Delhi | 185,000 | | 16,000 |
| Uttar Pradesh & Uttaranchal | 1,100,000 | | 20,000 |
| Bihar & Jharkhand | | 400,000 | |
| Rajasthan | 403,000 | | 330,000 |
| Madhya Pradesh & Chhattisgarh | | 242,000 | |
| Gujarat | 388,000 | | 399,000 |
| Maharashtra | 464,000 | | 70,000 |
| Goa | | | 18,000 |
| Karnataka | 118,000 | | 866,000 |
| Andhra Pradesh | 530,000 | | 281,000 |
| Tamilnadu | | | 204,000 |
| Orissa | | | 404,000 |
| West Bengal | | | 821,000 |

It is noticed that over-irrigated but poorly drained cultivated fields have suffered most from the scourge of alkalinization and salinization (Figure 4.9). The extent of salt-affected soils is considerable: 7–26 m ha (according to various researches). Of this, 2.5–3.6 m ha of cultivable land in U.P., Panjab, Haryana, Rajasthan and Gujarat suffer from alkalinization (Ghassemi et al, 1995).

## TREATMENT OF SICK SOIL

Since soil sickness is the result of over-irrigation in poorly drained areas, the most important and effective step in its treatment is to prevent seepage of water from irrigation channels. This can be accomplished by lining canals and waterways, and by building networks of drains to remove excess water. The drains need to be kept clear of silt and weeds so that there is no impediment to the smooth flow of water. These measures will lower the water table to below the critical level of 1.5 m below the surface.

The most common measure in the treatment of alkaline soil is to apply powdered gypsum to the upper layer (10–15 cm) of the alkali-affected soil. Adding sulphate of calcium (gypsum) at the rate of 2–15 tonnes/ha will convert the alkalis in the soil into soluble compounds, that can be flushed out by flooding the fields. Amendments like rice husk and straw, sulphatic press mud, phosphogypsum, zinc sulphate, farmyard manure, green manure, compost, etc expedite the recovery of alkali-affected soils. The application of low-grade pyrite—especially in the calcareous sodic soils of Bihar and eastern U.P.—has given encouraging results.

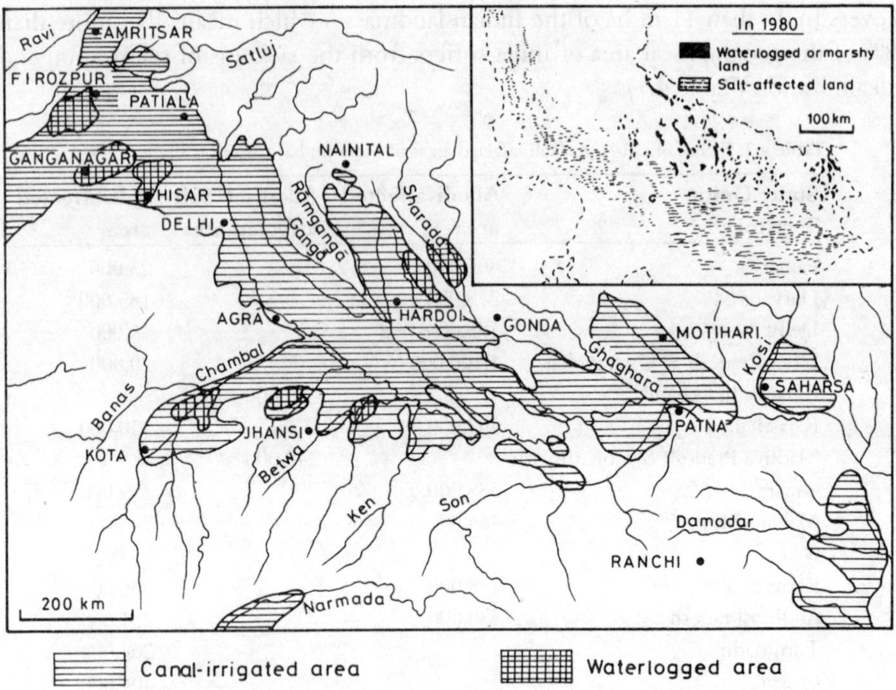


**Figure 4.9.** Canal-irrigated cultivated fields in northern India. The inset shows parts of the over-irrigated region that suffered (as in 1980) from salinization and water-logging. The situation today is much worse. (Based partly on Plate No. 190, *National Atlas of India*, National Atlas and Thematic Maps Organization, Kolkata, 1986)

Another step in combating the scourge of alkalinization and salinization is to plant salt-resistant vegetation—including barley, bersheem, pearl millet, raya, cotton, sugar-beet, date-palm, etc. Rice is known to hasten the reclamation of alkaline soil. Nearly 50,000 ha of abandoned salt-affected land in the Indo-Gangetic Plain has been rehabilitated by planting *Acacia nilotica, Eucalyptus tereticornis, Prosopis juliflora* (mesquite), *Terminalia arjuna*, *Albizzia lebbek* and the killer grass *Leptochloa fusca*.

## Losing Ground

As already shown, the loss of vegetation triggers deterioration and denudation of the soil. Incapable of withstanding multiple assaults, much of the land in India is in serious distress. More than 53% of the geographical area (~155 m ha) of India (Figure 4.10, Table 4.2) is suffering from erosion—with more than 70 m ha in critical condition (Babu et al, 1977). According to a recent estimate made by the Tata Energy Research Institute, over 200 m ha of land would be degraded through erosion by 1997. The erosion is particularly pronounced in areas where the landscape is being reshaped recklessly, and where the protective shield of forests (Plate 2) is being damaged. Nearly 60% of agricultural fields are vulnerable to erosion and its attendant degradation. Every year, 5–7 m ha of farmland falls victim to erosion of one kind or another. The rate of loss of soil per square kilometre area annually is as follows: northwestern Himalaya 15.93 tonnes/km$^2$, Panjab–Haryana alluvial plains

330 tonnes/km$^2$, upper Gangetic plains 1410 tonnes/km$^2$, lower Gangetic plains 940 tonnes/km$^2$, and northeastern Himalaya 50 tonnes/km$^2$ in the forested region, and 4095 tonnes/km$^2$ in areas of the *jhum* cultivation (Dhruva Narayan, 1993).

Significantly, along with the erosion-generated detritus go the valuable nutrients that the soils contain. It takes centuries for the soil to build up this store of nutrients—which is washed away in what is comparatively the blink of an eye, through denudation and erosion. The topsoil containing nutrients is eroded away at the rate of 3 million tonnes per year (2.5 m t/yr in 1950). It is estimated that 5.37–8.4 million tonnes of nutrients in the soil are lost every year (Babu et al, 1977). Obviously, this is the loss of a valuable resource.

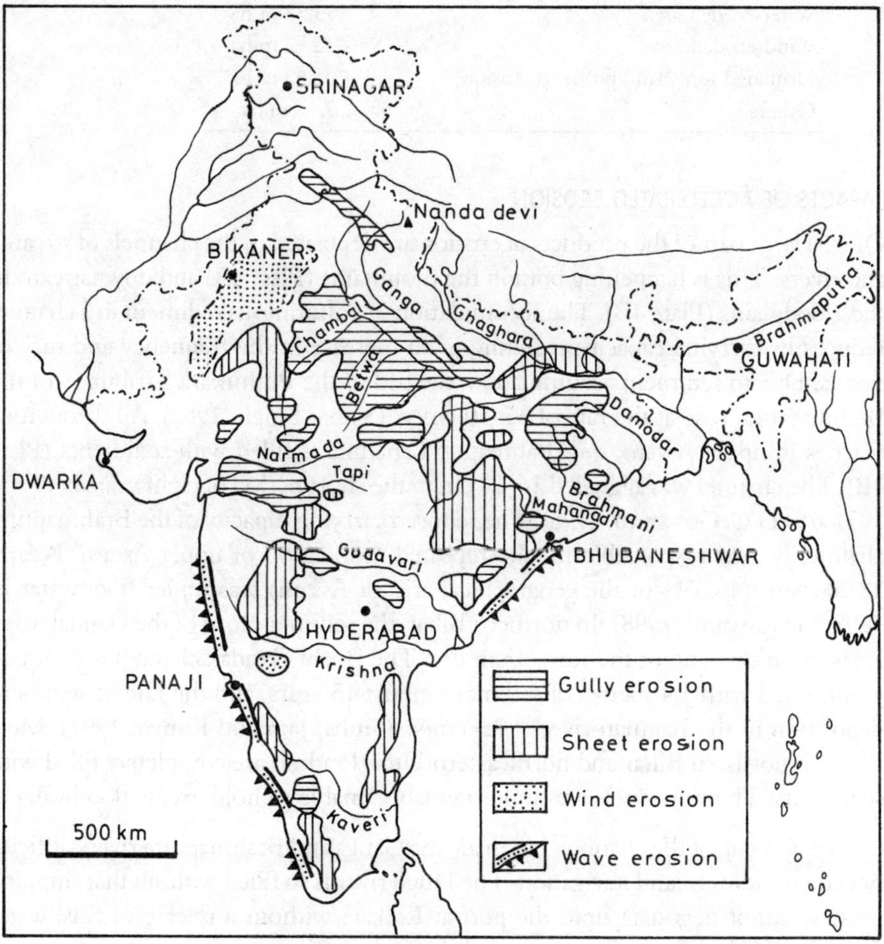


**Figure 4.10.** Areas affected by erosion of one kind or another.
(Based on Kaul and Das, 1982).

The magnitude of this **soil-loss** can be judged from the amount of sediments the rivers are carrying to the sea. The aggregate of sediment discharged by Indian rivers is 1,490 million tonnes/yr (Subramanian and Dalvi, 1978). The Ganga alone transfers upwards of 411 million tonnes of sediment annually to the Bay of Bengal (Abbas and Subramanian, 1984). The Brahmaputra carries 402–710 million tonnes of sediments annually (Subramanian and Ramanathan, 1996).

**Gully erosion** has converted 3.97 m ha of land in India into ravine land or **badland** (Plate 3). The gullies have swallowed valuable land at the rate of 8000 ha/yr (Dhruva Narayan, 1993; Babu et al, 1977). Spectacular ravine lands are seen in the Chambal–Yamuna valleys in the U.P.–M.P. area (1.91 m ha), in the Mahi valley in Gujarat (0.45 m ha), and in the Damodar basin in Jharkhand–West Bengal (Abrol, 1994).

**Table 4.2.** Extent of eroded land in India (Abrol, 1994).

| | |
|---|---|
| Total wasteland area | 129.57 m ha |
| Water-eroded area | 73.58 m ha |
| Wind-eroded area | 12.92 m ha |
| Degraded forestland prone to erosion | 35.85 m ha |
| Others | 7.16 m ha |

## IMPACTS OF ACCELERATED EROSION

Quite a large part of the products of erosion are deposited in the channels of streams and rivers. This is happening both in the mountains (Plate 4A) and downstream in the floodplains (Plate 4B). The accumulation of voluminous sediment in a channel reduces its carrying capacity, resulting in an increase in the frequency and fury of floods. Due to sediment accumulation, the bed of the Puthimari, a tributary of the Brahmaputra, rose at the rate of 35–50 cm/yr (Samar Singh, 1982). All through its reaches in upper Assam, the Brahmaputra channel is filled with sediments (Plate 4B). The channel was aggraded 16–21 cm in the short period of eight years between 1971–79 (D.C. Goswami, 1998). The reduced carrying capacity of the Brahmaputra channel is thus responsible for the repeated inundation of upper Assam. Nearly 38,200 km$^2$ (46.65% of the geographical area of Assam) was under floodwater in 1998 (S. Goswami, 1998). In northern Bihar, the sediment load of the Gandak river is 79–80 m t/yr, and of the Kosi, 43 m t/yr. The Burhi Gandak channel was, at one point, filled with 1.4 metres of sediment in just 15 years, and the rate of sediment deposition in the Bagmati river is 9.3 cm/yr (Sinha, Jain and Kumar, 1997). Most rivers in northern Bihar and northeastern Uttar Pradesh are completely filled with sediments. Their channels are understandably unable to hold excess floodwater.

The shallowing of the channels has made the Ganga and Brahmaputra rivers difficult or even unfit for inland navigation. The Hugli river is so filled with silt that shipping vessels cannot negotiate upto the port at Kolkata without a release of flushwater from the Farakka Barrage.

Sizeable proportions of the eroded material are trapped by natural lakes and artificial reservoirs behind dams. Taking into consideration the data related to various river valley projects in different parts of India, it is estimated that the reservoirs are accumulating sediments at the rate of 2.5–18 ha-m/km$^2$/yr (Gupta 1975; Lagwankar et al, 1994). This has considerably shortened the useful lifespan of these reservoirs, lessening the benefits of water resources development projects.

## Soil Conservation Measures

With their knowledge of the dynamics and mechanisms of earth processes, geologists are well-equipped to develop an appropriate and efficient strategy to protect soils. While an agricultural scientist endeavours to restore the productivity of the degraded soils, the geologist is more concerned with controlling erosion of the land in general. These erosion-controlling measures aim at not letting water and wind sweep away the soils. This can be achieved by: (i) preventing dislocation of the slope-forming material, (ii) draining out runoff water before it begins to erode, and (iii) planting a protective cover of vegetation on vulnerable spots or wounded land. In other words, the integrity of the ground needs to be ensured through combined biological and engineering measures. The package of measures, which may vary from situation to situation, constitute what is known as a **Watershed Management Programme** (Figure 4.11).

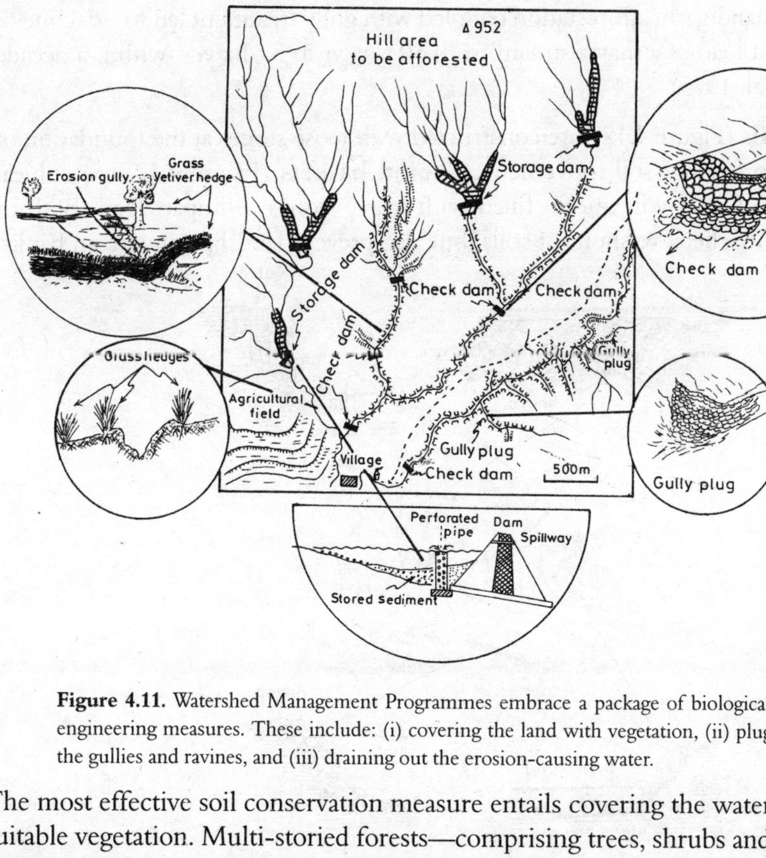


**Figure 4.11.** Watershed Management Programmes embrace a package of biological and engineering measures. These include: (i) covering the land with vegetation, (ii) plugging the gullies and ravines, and (iii) draining out the erosion-causing water.

The most effective soil conservation measure entails covering the watershed with suitable vegetation. Multi-storied forests—comprising trees, shrubs and grasses—provide effective insurance against denudation and resultant soil-loss. Plants that grow quickly and have abundant foliage and deep roots make the best choice.

The plants are chosen on the basis of the agro-climatic zone in which the soil-treatment is to be carried out. Vetiver grass (*Vetiveria zizanoides*) is being extensively used to protect stream banks, highway slopes (Plate 5) and mined areas. A quick-growing grass, with deep root networks, it grows into a dense hedge in three to four months' time and protects the slope from eroding and slipping (Figure 4.11 and Plate 5). Low in cost, vetiver has high tolerance to a variety of toxic conditions resulting from pollution.

Where the land has suffered deep gully erosion such as in the badlands of the Damodar, Chambal and Mahi valleys, watershed management requires plugging of gullies and building check dams at close intervals across ravines (Figure 4.11). Using stakes of self-generating trees (like poplars and willows) in the construction of check dams ensures their stability. Appropriate sites are chosen for constructing check dams after taking into consideration the factors and the features that cause leakage of impounded water. To give an example of a successful case, the afforestation in the catchments in the Birahi, Garurganga and Patalganga tributaries of the Alaknanda river (in Garhwal) resulted in the reduction of active landslide area from 1,409 ha in 1972 to 1,093 ha in 1995 (Kimothi and Juyal, 1996). In the Sukhomajri (Siwalik) area near Chandigarh, afforestation coupled with gully treatment led to a decline of the sediment load of a major stream from 50 t/ha/yr to 5 t/ha/yr—within a decade (Tejwani et al, 1975).

**Riprap walls** (Figure 4.12) are constructed with loose stones at the foundation of structures to prevent soil movement. **Gabion baskets** (Figure 4.12) are cubical cages made of steel-wire mesh, filled with heavy stones to impart load. Both of these not only check washout of soils, but also prevent landslips on stream banks.

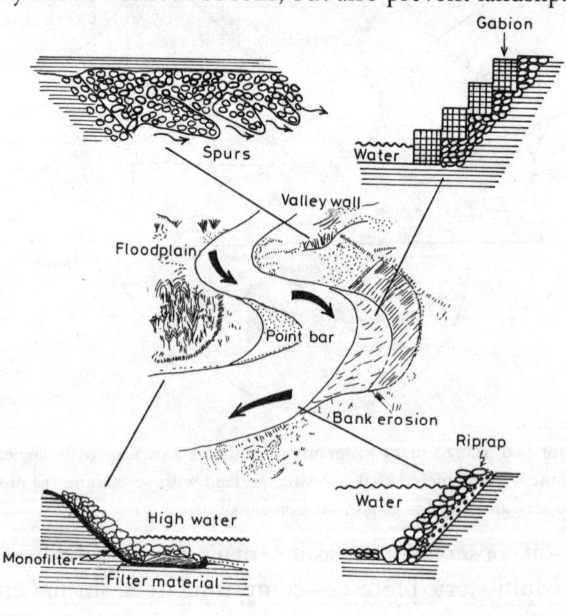


**Figure 4.12.** Spurs, gabion baskets and riprap walls protect banks and embankments against lateral erosion in stream valleys. The spurs deflect currents and reduce their abrasive power.

**Spurs** deflect currents and lessen their velocity and violence. Mindful of the dynamics of stream flow, geologists are able to suggest appropriate locations to erect these structures.

It may be mentioned that bank erosion has assumed serious proportions in the floodplains of such rivers as the Kosi and the Gandak in northern Bihar, and the Brahmaputra in Assam and Bangladesh. During floods, these rivers tend to shift their courses, cutting their banks and enlarging the loops. Agricultural land along the banks is seriously threatened. In Bangladesh, the rate of stream migration being more than 1,000 m/yr, the banks of the Jamuna (the Bangladeshi course of the Brahmaputra) are eroded at an average rate of 50 m/yr, the maximum being 400 m/yr (Barua, 1997).

To combat erosion of dunes along the sandy coasts and in dry deserts, the conservation measures involve construction of features that would reduce the velocity of winds. For this purpose, multiple rows or fences of trees and shrubs are developed across the paths of winds. These are called **wind breakers** and 'shelter belts' (Figure 4.13). The trees are planted close enough to form a continuous barrier. Application of chemicals, asphalt and rubber solution tend to stabilize the sand which accumulates in front of the plant barriers. Interestingly, wind breakers were first built in Suratgarh district in Rajasthan when large farms were developed in the dry domain.

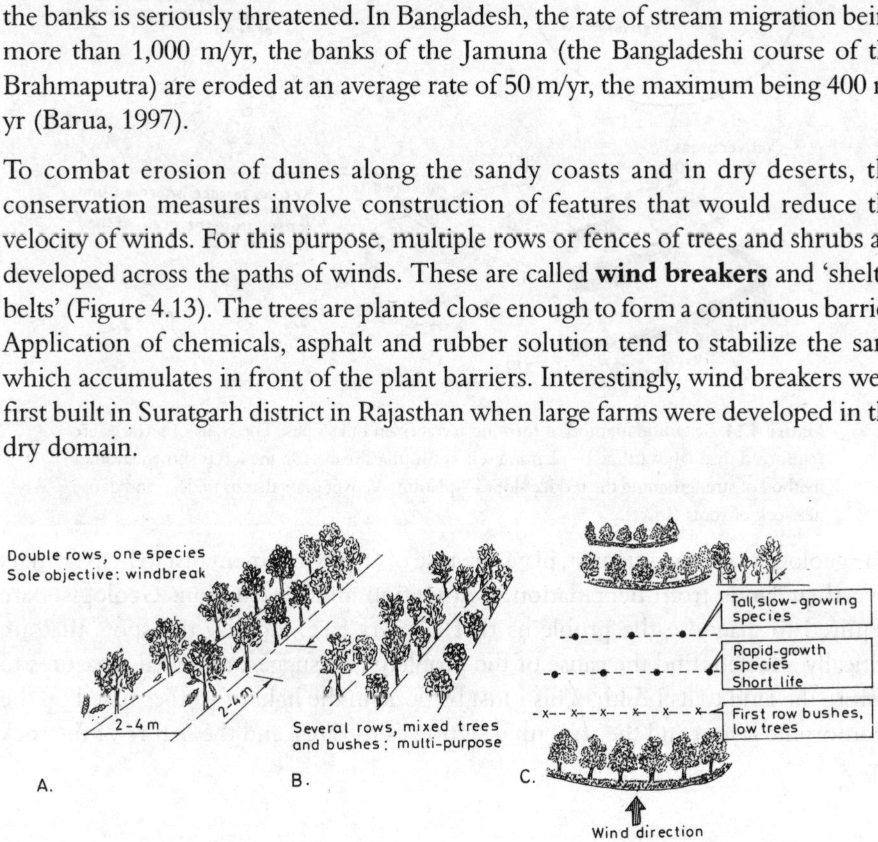


**Figure 4.13.** Shelter belts serve as wind breakers and check the march of sand dunes in deserts and along sandy coasts.

## HELPING FARMERS IN HILLY REGIONS

Preservation of the nutrient-charged topsoil is the supreme concern of farmers. In hills and mountains, this is an important matter to help sustain farming, and in turn, the survival of the people. A scientific method of saving the soil is to cut the slopes into terraces. These terraces must be parallel or conformable to the contour lines. They need to be reinforced through **wattling** of quick-growing or self-sprouting shrubs (Plate 5, Figure 4.14) or trees planted at suitable intervals. The excess rainwater must be drained out through shallow furrows dug into the inner side of the terraces, and by making low levees on the outer margins.

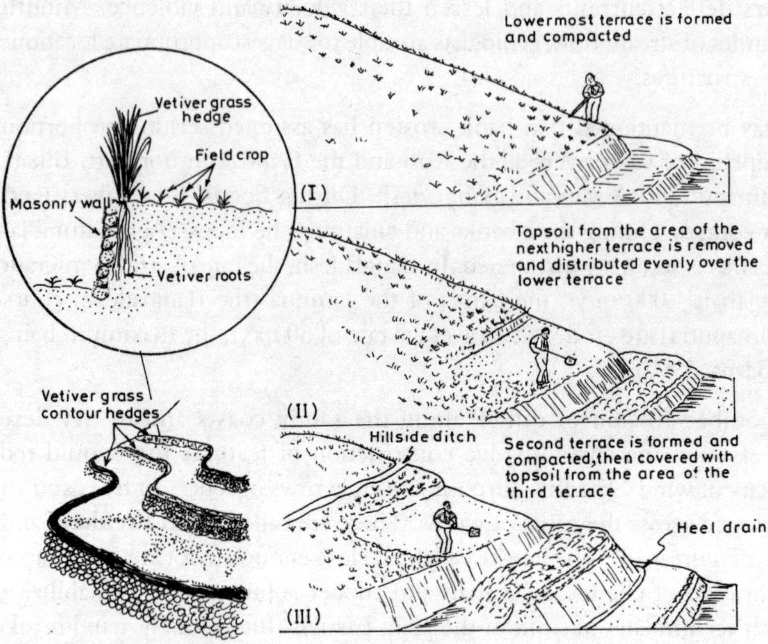


**Figure 4.14.** Scientific method of forming terraces on hill slopes. The walls of terraces are reinforced through wattling by planting self-sprouting shrubs. On the left is shown another method of strengthening the terrace slopes—planting Vetiver grass that has a deep and dense network of roots (Inset).

The geologist, as a soil scientist, plays the role of an environmentalist to help farmers save their lands from degradation, denudation and destruction. Geologists are required to diagnose the problems that farmers face, identify the spots that are critically affected, find the cause of the problem and suggest remedial measures to restore the land to its health. This must be done in the light of studies based on the geomorphic layout and the structural setting of the area, and the nature of the rock types.

# 5

# Caring for Water Resources

## Water: A Replenishable Resource

Water is a finite resource, and quite scarce in most parts of India. In the global context, freshwater—flowing down rivers, held stagnant in lakes, and lying frozen as snow and icepacks—accounts for about 2.7% of the earth's total water resources. And this, despite the fact that nearly 71% of the earth's surface is covered with water!

In India, **total rainwater resources** amount to nearly 400 million hectare-metre per annum. Of this, about 185 m ha-m is available as surface flow, 50 m ha-m is stored underground and 165 m ha-m rests in soils, part of which is lost due to evaporation. Indian rivers carry nearly 1,645,000 million $m^3$ of water, every year, to the Indian Ocean (Rao, 1975). The freshwater reserve held in ponds, lakes and reservoirs, and stored underground is 431,423 million $m^3$—groundwater alone amounting to about 255,000 million $m^3$ (Kayastha, 2000). Replenishable groundwater resources are estimated by the Central Groundwater Authority at 431,320 million $m^3$, out of which the utilizable amount is 395,600 million $m^3$ (Ghosh and Bathija, 2001). Another estimation shows that only a little more than 45,000 million $m^3$ is utilized (Kayastha, 2000). It is evident that water resources are not being fully tapped, or not used gainfully or appropriately.

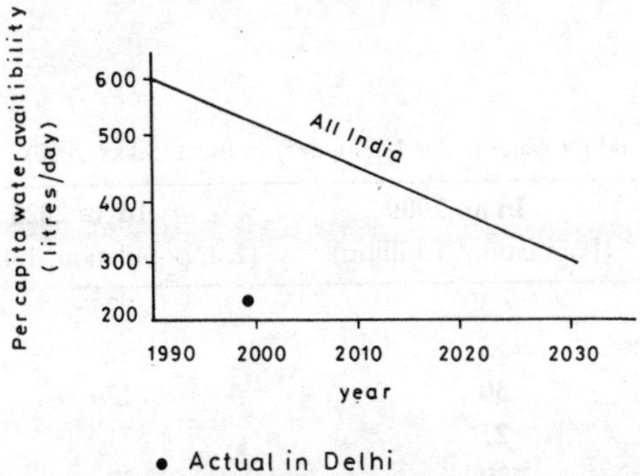


**Figure 5.1.** Projected per capita availability of water in India as a whole. Note that the actual amount available in the capital of India is far less than the requisite quantity (Subramanian, 2001).

An overwhelmingly large (>85%) part of rainwater in India is wasted in floods. This is because the monsoon here is restricted to only a few months in a year—and the precipitation occurs in spurts, especially during storms. In practical terms, the total duration of rain in India is restricted to merely 300 hours in a year (Pisharoty, 1987).

## THE GROWING NEED FOR WATER

Against this scenario of abundant water in rivers and underground aquifers, little more than 19 per cent of Indian villages get water from *reliably safe* supply systems. About 8,000 villages have no water at all. The **per capita water consumption** in Indian towns is less than 90 litres/day, although in big cities (Figure 5.1) it is as high as 214 l/d (Kayastha, 2000; Agarwal et al, 1999). In other words, the average water consumption in India—where a water supply system exists and is functioning—is about 205 l/d. Compare these figures (Table 5.1) with the per capita water consumption in the United States of America, which is about 530 l/d!

**Table 5.1.** Per capita water consumption (in litres/day) in some major cities in India (Agarwal et al, 1999).

| | | | |
|---|---|---|---|
| Ahmadabad | 182 | Bangalore | 137 |
| Bhopal | 234 | Coimbatore | 104 |
| Delhi | 341 | Chennai | 87 |
| Hyderabad | 127 | Indore | 208 |
| Jaipur | 195 | Kanpur | 200 |
| Kochi | 231 | Kolkata | 200 |
| Lucknow | 252 | Ludhiana | 175 |
| Madurai | 74 | Mumbai | 272 |
| Nagpur | 158 | Patna | 297 |
| Pune | 241 | Surat | 178 |
| Vadodara | 233 | Varanasi | 215 |
| Visakhapatnam | 113 | Average megacities | 214 |

**Table 5.2.** Annual demand for water (cubic kilometres) in India (Sikka, 2002).

| Purpose | In AD 2000 (Population: 1 billion) | In AD 2025 (Est. population: 1.4 billion) |
|---|---|---|
| Irrigation | 630 | 777 |
| Livestock and domestic | 33 | 52 |
| Industries | 30 | 120 |
| Thermal power plants | 27 | 71 |
| Miscellaneous | 30 | 37 |
| Total | 750 km$^3$ | 1,050 km$^3$ |

Water requirement in India is very high (Table 5.2). Towards the end of the year 2001, the total demand for water was about 13,700 million litres per day. At the present rate of consumption, India will require 140% more water by AD 2025 (Table 5.2). Clearly, new reserves of water must be found and all existing reserves augmented by every method known. This implies that geologists—as geohydrologists—have to play an increasingly active and crucial role as the finders and conservators of water resources. Once it is understood that the amount of usable freshwater is limited, humans will ensure that this 'elixir of life' is used judiciously, and all efforts will be made to conserve and replenish its dwindling resources.

## Availability of Riverwater

As already stated, the rivers of India (Plate 6) discharge freshwater at the rate of 1,645,000 million $m^3$/yr into the Indian Ocean (or 1,853,000 million $m^3$ according to Agarwal et al, 1999). More than 85% of this discharge is carried by major rivers, 7% by minor rivers and 8% by ephemeral streams. The rivers are fed by countless springs and seepages that draw out water stored in underground aquifers. Spring water is supplemented by snow-meltwater, which amounts to 800 $km^3$ per year. Nearly 49,200 $km^2$ of the Himalayan region is covered by icepacks and glaciers, and 50–70% of the water flowing in the rivers of the northwestern Himalaya is meltwater, while the rivers of the eastern Himalaya carry less than 30% meltwater. The Satluj (with 227 glaciers) and the Chenab (with 200 glaciers) contribute significant amounts of meltwater—the average **meltwater** contribution to the annual flow of the Chenab is 49.1%.

As a result, the occurrence of floods, and the annual discharge, increase perceptibly during pre-monsoon months, when the surface temperature rises (Bhutiyani et al, 2002). Since the melting of snow coincides with the monsoon—when the river discharge is already extremely high—the meltwater is lost in floods. Precious meltwater is thus totally wasted.

While the melting of glacial ice supplies water to the Himalaya-born rivers, it also implies the retreat of glaciers. This has serious implications. For example, the Gangotri glacier retreated at the rate of 14.6 m/yr in the period 1935–76; 46.4 m/yr during 1962–90; 37.5 m/yr in the interval 1990–94; and 25 m/yr in the period 1994–98 (Tangri, 2000). All other glaciers have retreated in a similar manner.

If glaciers continue to retreat at this rate, a time may come when there will be no ice left to generate meltwater. One can imagine the scenario when there is no meltwater contribution to these rivers (Table 5.3). The water that flows during the rainy season in the snow-fed Karnali river in Nepal is at least one thousand times more than its discharge during the dry season (Kattleman, 1987). Obviously these rivers have high discharge during the three to four months of the monsoon, but become petty rivers with low discharge in the remaining eight to nine months of dry weather—making the eventual withdrawal of meltwater contribution a grave and ominous consequence.

**Table 5.3.** Average annual flow of rivers of the Indian subcontinent (Based on Rao, 1975; Agarwal et al, 1999)

| River (measuring point) | Flow (million $m^3$/yr) |
|---|---|
| **Brahmaputra System** (Barak) | **537,100** |
| Brahmaputra (Guwahati) | 455,000 |
| Manas (Mathanguri) | 47,100 |
| Others | 35,000 |
| **Ganga System** (Farakka) | **501,600** |
| Ganga (Hardwar) | 23,900 |
| Yamuna (Saharanpur) | 12,300 |
| Gandak (Patna) | 53,200 |
| Tista (Anderson Bridge) | 19,400 |
| Others | 392,800 |
| **Sindhu System** | **207,800** |
| Beas | 15,800 |
| Satluj | 16,700 |
| Ravi | 8,000 |
| Chenab | 29,000 |
| Jhelam | 27,900 |
| Sindhu | 73,300 |
| **Others** | **37,100** |
| Ghaggar | 1,300 |
| Madha | 2,500 |
| Tupan | 4,000 |
| Luni | 8,000 |
| Banas | 8,700 |
| Sabarmati | 3,800 |
| Mahi | 8,500 |
| Narmada | 40,900 |
| Tapi | 18,000 |
| Godavari | 119,000 |
| Krishna | 67,800 |
| Pennar | 6,900 |
| Kaveri | 21,000 |
| Subarnrekha | 10,800 |
| Brahmani | 36,300 |
| Mahanadi | 66,900 |
| Barak and others | 59,800 |
| **Total** | **1,853,000** m $m^3$/yr |

## POLLUTION OF RIVERS: HOW MUCH ARE HUMAN BEINGS RESPONSIBLE?

The fourteen major rivers of India meet nearly 85 per cent of the country's water needs. Some of these rivers have been revered in ancient times, as articulated in the following *shloka* that is recited before Hindu religious rites:

गंगे व यमुने चैव गोदावरि, सरस्वति, नमदे,
सिन्धो, कावेरि जलेऽस्मिन सन्निधिं कुरू ।

Oh rivers Ganga, Yamuna, Godavari, Saraswati, Narmada,
Sindhu and Kaveri! Come and fill this pot before I start this rite.

Ironically, all fourteen rivers are now greatly polluted. Of the 44,000-km aggregate length of all the rivers in India, more than 6,000 km are unfit to drink, as the dissolved oxygen content is much below the required 3 milligrams per litre.

Minor rivers and streams passing through cities and towns are so filled with pollutants that they appear to be no better than gutters. Even major rivers like the Ganga, Yamuna and Sabarmati—flowing past megacities—have begun to resemble waste-drainage systems, characterized by heavy loads of organic waste, micro-organisms and toxic chemicals released by neighbouring industries. A consequence of this unchecked discharge of untreated municipal and industrial effluents is that the **dissolved oxygen content** in riverwater has gone down much below the critical level of 3–4 mg/l. Toxic elements like lead, zinc, chromium, copper, cadmium, etc, have even entered into the framework of the clays of riverbed sediments. The Yamuna continues to function as a 'drain', with 490 km of its course being nearly dead. The 22-km course it runs through Delhi contributes to 80% of its pollution. This is understandable, for the bustling capital generates sewage at the rate of 3,300 million litres per day, of which only 1,500 million l/d is treated, while 1,800 million l/d flows into the Yamuna (*Down to Earth*, 30 April 2003, p. 16).

These toxic elements, in trace amounts, have found their way into the food chain of animals and humans—who consume the polluted water and the grasses that grow on the floodplains of the polluted rivers. This is because the floodwaters carry clays that have absorbed toxic elements from industrial effluents, and spread this onto the plains supporting fodder grasses. On the whole, riverwater is becoming less and less consumable thanks to the phenomenal increase in pollution everywhere in the country. All present attempts to reduce pollution have met with disastrous failure.

## Rainwater Infiltration: Storage Underground

Only a part of the rainwater is absorbed by the ground—it depends entirely on the land being permeable enough to allow the passage of water. The infiltration of water depends on:

(i) Texture of soil or structure of the bedrock,

(ii) Extent and nature of vegetation that covers the ground,

(iii) Degree of ground slope, and

(iv) Intensity and duration of rainfall. On average, only 15% of the rainfall is able to penetrate and replenish aquifers—the range of infiltration being 3–25% (Figure 5.3).

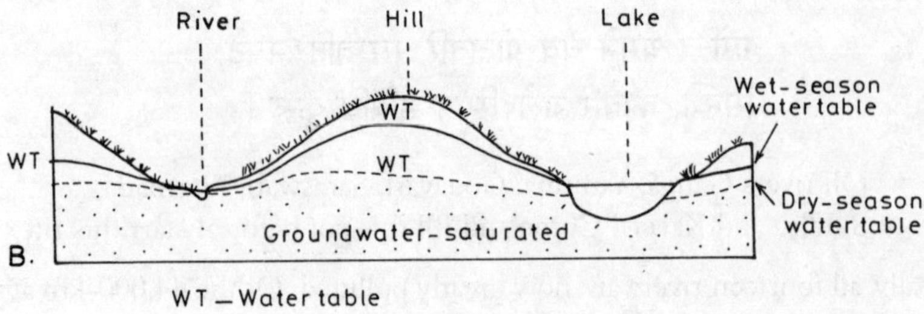


**Figure 5.2.** Infiltrating water fills pores and openings within soils and bedrocks, and forms a zone of saturation. The upper limit of this water-saturated zone is called the water table (WT). When the water table is cut by the topographic surface, springs and seepages originate and contribute water to rivers and lakes.

The infiltrating water fills pores and openings in the soil (Chapter 4, Figure 4.2 *inset*) and in the bedrock below the soil, to form a **zone of saturation** in the deeper layers (Figure 5.2). The upper limit of the zone of saturation is called the **water table**. Above the water table, the pores are filled with air and moisture. This is the **zone of aeration**. Between this zone and the saturation zone lies the **capillary fringe**, about 1–2 cm in thickness. The sustenance of vegetation depends on the zone of aeration and the zone of saturation, but particularly on the former (Figure 4.2 *inset*).

The downward percolation of water depends on a number of factors, as stated above. The coarser the texture of the soil and the more fractured the rocks, the greater would be the infiltration. Additionally, vegetation and/or plant litter not only protect the soil from the consistent pounding of raindrops, but also provide the organic matter (humus) that makes the soil porous and permeable. The permeability allows interstitial water to move from higher to lower levels. The elimination of the vegetal cover therefore considerably reduces the rate of infiltration (Figures 4.6 and 5.3).

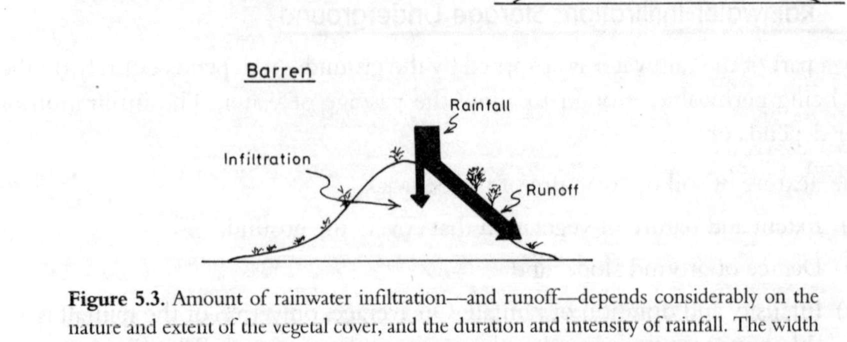


**Figure 5.3.** Amount of rainwater infiltration—and runoff—depends considerably on the nature and extent of the vegetal cover, and the duration and intensity of rainfall. The width of the arrow is indicative of the proportion of water infiltrating the land or escaping as surface flow or runoff.

Likewise, the absence or removal of weathered rocks, or the covering of rocks and soils with impermeable material (like concrete or coal tar), prevents the ingress of water. Intense rainfall loosens the soil particles and causes the pores to get clogged, thus inhibiting rainwater infiltration. Moreover, the rapidity of heavy rainfall does not permit easy percolation. On the other hand, slow rainfall over long periods is conducive to better recharge of groundwater. To sum up, the soil/bedrock is able to absorb larger amounts of water if the rainfall is gentle and prolonged, particularly where the ground is flatter or gently sloping, if the surface is covered with a carpet of vegetation and litter (Figure 5.3), and if the soil is open-textured (porous or fractured).

The context of water infiltration and **groundwater recharge** can be used to view the loss of forest cover in India (Plate 2). Natural forests today cover not more than 19% of the geographic area of the country. Much of these forests are in a degraded condition, dense forests covering only 1.73% of India. This alarming situation—where 81% of the country's land surface is bereft of forests—has had considerable impact on the recharge of groundwater through rainwater percolation.

### THE WATER TABLE AND INVISIBLE UNDERGROUND FLOW

As already stated, the upper surface of the groundwater is called the **water table**. The water table conforms roughly to the topography of the land (Figure 5.2). Its depth depends on a number of factors. In areas of heavy to moderate rainfall, but with gentle slopes, it may be within a few metres of the ground surface. In rainless regions, the depth of the water table can be as much as 50 m or more. In forested areas, where the soil texture is coarse, the water table is quite near the surface; but deep below if the texture is fine and the terrain lacks the cover of vegetation.

The water table fluctuates conspicuously from season to season and situation to situation. Times of drought, and heavy human consumption (using pumps and wells) lead to withdrawal of groundwater, and the water table sinks to deeper levels. The rate of fluctuation depends on the rate of movement of the groundwater, varying from a few centimetres to tens of metres per day. This movement is governed by the **permeability** of the medium—its ability to allow passage of water without impairment of the structure or fabric of the medium. The permeability in turn depends on the size of the pores and openings, and their interconnection. In coarser, permeable sands, or in highly fractured/jointed rocks, the movement is many times quicker than in soils with fine-textured sediment, or massive rocks that have few fractures. There is practically no movement of groundwater in soils that are rich in clays. This situation culminates in the accumulation of water in pores and other openings, causing **waterlogging** of the land.

The direction of the flow of groundwater is of crucial importance in locating well sites and for monitoring trends of groundwater pollution. The geologist depends on knowledge of geomorphology, sedimentology, structural geology and geohydrology to delineate the movement trends of invisible underground water—

to locate its position, temporary or permanent residence, and ways of draining it out moderately (Figure 5.4).

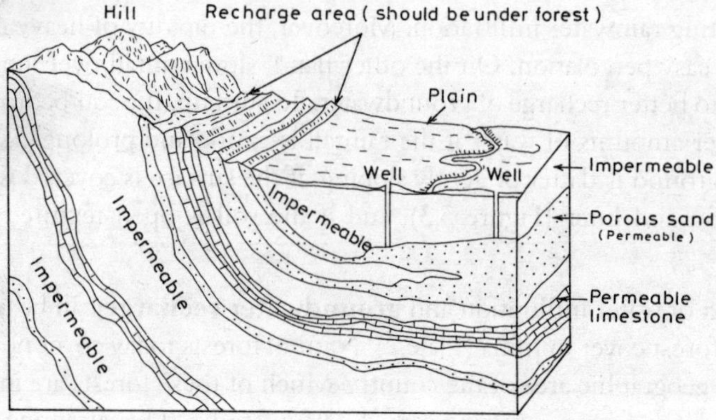


**Figure 5.4.** It is the geologist's job to identify and delineate the area where the surface and rainwater percolates underground (the recharge area) trace the flow of underground water through the medium (sediments or rocks), locate its place of residence, and the options of drawing it or draining it out.

## Groundwater Resources

Porous-permeable zones, including weathered and/or fractured-jointed rocks—where interstitial water is able to move freely and can be tapped for use—are called **aquifers** (Figure 5.5). The capacity of an aquifer for holding water depends on its porosity, which is the volume of empty space between the grains that constitute sediment/soil (Figure 5.6A), or structural discontinuities like fissures, fractures, joints and faults in the rocks (Figure 5.6C). The ability of the soils or rocks to allow the passage of water, without impairment, is called permeability. Permeability depends on the size of the pores and fractures and their interconnection. In coarser sands or soils or fractured rocks, the movement of water is many times faster than in fine-textured, less permeable material. In clay-rich soils there is practically no movement of groundwater.

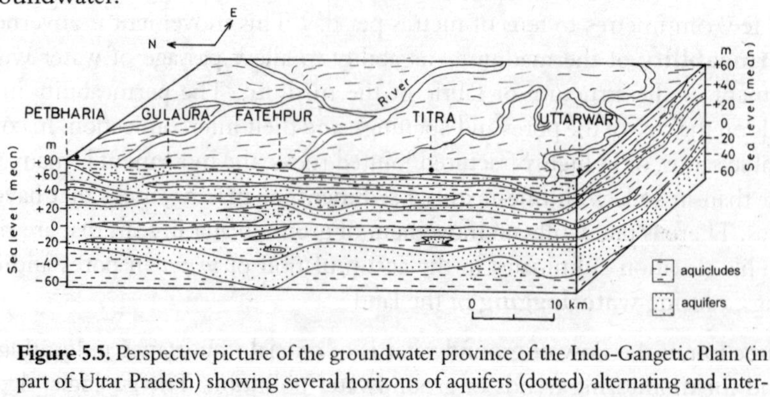


**Figure 5.5.** Perspective picture of the groundwater province of the Indo-Gangetic Plain (in part of Uttar Pradesh) showing several horizons of aquifers (dotted) alternating and inter-fingering with aquicludes (broken lines). Note the perched water pockets within the aquicludes. The aquifers are made up of sand, local gravel and *kankar*. The aquicludes constitute mud and clay.

Shale, mudstone and other dense impermeable rocks are called **aquicludes** (Figure 5.5). Aquicludes within permeable aquifers impede percolation of underground water, and give rise to what is called **perched water**. Perched water is the unconfined groundwater that is separated from the main body of groundwater below by a zone of unsaturated sediment/soil. Accumulation of iron oxides, salts and calcretes (concretionary carbonate nodules) in layers, or as hard pans within the soil often results in the development of shallow perched water.

Covering an aggregate area of 905,000 km², the alluvial plains of the Sindhu, Ganga, Brahmaputra, Mahanadi, Godavari, Krishna, Pennar, Kaveri, Tapi, Narmada, Mahi and Sabarmati river systems provide an abundance of aquifers. The variably thick successions of fluvial sediments constitute the richest groundwater province of India. The reserves are estimated at 52.52 million ha-m in the Ganga basin, 53.74 million ha-m in the Brahmaputra domain, and 5.98 million ha-m in the Barak region. In the Ganga basin, the aquifers vary from less than 20 m to as much as 330 m in thickness. The sustained yield of tubewells varies from 4,170–25,000 l/minute at the drawdown of 3–4 m (Pathak, 1983). In the eastern part of the Ghaghara–Gandak interfluve area, where the regional water table lies at a depth of 3–5 m, two major aquifers upto the depth of 130 m have a yield of 1,200–2,850 m³/day. Paleochannels, represented by oxbow lakes in the Indo-Gangetic Plain, have great potential of sustained water supply.

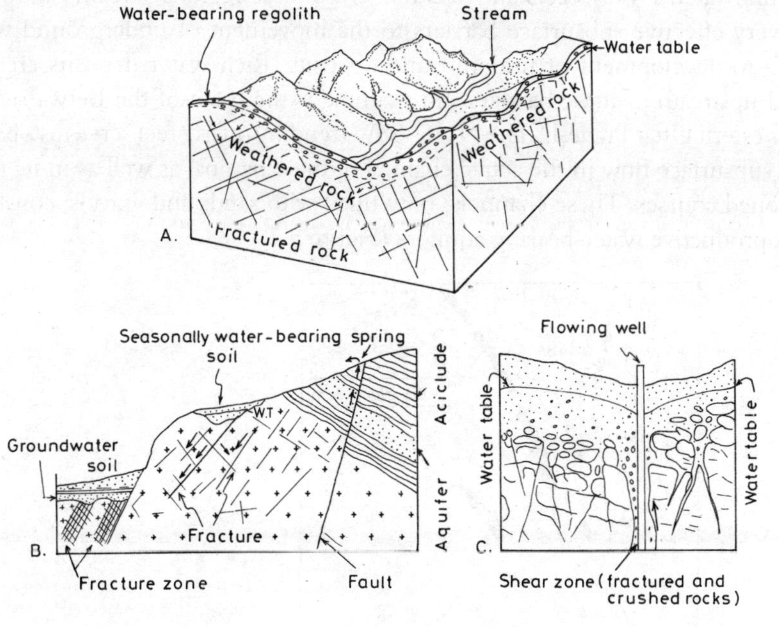


**Figure 5.6.** (A) Weathered rocks above fractured and jointed bedrocks are a very favourable location for groundwater. Streams are fed by seepages related to the aquifer.
(B) Shallow fractured rocks and zone of weathering serve as good aquifers. The water collects in depressions.
(C) Shear zones and associated zones of crushed and weathered rocks are extremely good repositories of groundwater.

Zones of rock-weathering are often filled with water. The thickness of the weathered zone can be as much as 5–15 m. In central and southern India, weathered rocks are capped by mantles of laterite or lateritic soil upto tens of metre in thickness. Weathered rocks are a very good repository of groundwater (Figure 5.6A). The weathered zone is often underlain by fractured or jointed rocks (Figure 5.6A and 5.6B). The zones of fractures, faults and shearing (close-spaced multiple faulting) store great volumes of water (Figure 5.6C). Remote sensing surveys have shown that the NE–SW and NNW–SSE trending faults in the Karnataka–Tamilnadu regions are associated with high-yielding productive wells, particularly at their intersections with the E–W to WNW–ESE oriented fractures (Srinivasan and Sreenivas, 1977; Ramasamy, 2002). Detailed geophysical prospecting helped to delineate the N50°E–S50°W trending shear zones in the northwestern part of Dindigul district (T.N.) which now supply very productive wells (Chandra et al, 2002). The deeply weathered top of Bundelkhand Granite in southwestern Uttar Pradesh, and Peninsular Gneiss in Andhra Pradesh and Karnataka, for example, make excellent aquifers. The tubewells of Jhansi district (U.P.) dug into the weathered rocks atop Bundelkhand Granite, yield abundant water, at rates varying from 1,000 to 100,600 l/min.

In the Malwa and Deccan plateaus, the highly jointed and locally vesicular lavas of basalt, with their associated volcanic ash beds and buried red soils (sandwiched between successive lava flows and often referred to as 'red boles') have good stores of groundwater. Dykes, reefs and sills that cross these regions' streams and rivers form very effective subsurface barriers to the movement of underground water, leading to development of underground deposits. Rich water deposits are thus located upstream of these barriers. An example is the basin of the Betwa river in southwestern Uttar Pradesh. The NNE–SSW trending quartz reef serves as a barrier to the subsurface flow in the active channel of the Chambal as well as in its many abandoned courses. These channels, now filled with sands and gravels, constitute highly productive water-bearing aquifers (Figure 5.7).

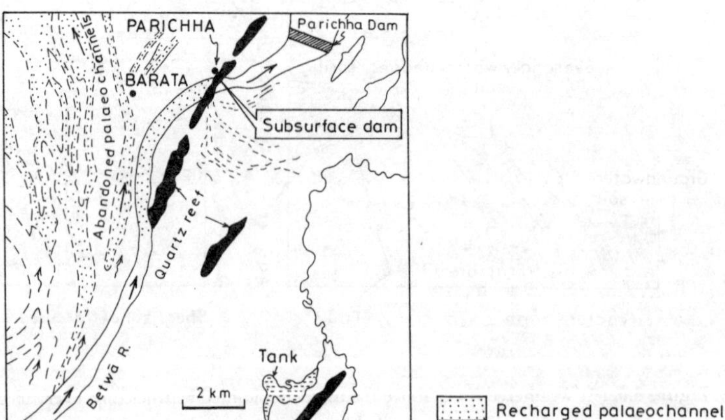


**Figure 5.7.** NNE–SSW trending quartz-reef serves as a barrier to underground flow in the channel of the Chambal river, as well as to the multiplicity of courses it abandoned in the geological past (After Chaturvedi et al, 1983).

Old abandoned channels of rivers in tectonically resurgent terrains in Peninsular India have been found to have rich repositories of freshwater (Figures 5.7 and 5.8). It is noticed that the sand-filled, wide, shallow channels of the Chambal, Betwa, Son, Indravati, Ghod, Mutha, etc have been found to be very productive—yielding ample quantities of water, even though on the surface they have very little dry-season discharge. In Peninsular India, the valley fills make good aquifers. Where the thickness of the valley fills is 5–10 m, infiltration galleries (Figure 5.22) can be used to tap subsurface flows. However, large valley fills (about a kilometre in width) do not carry enough subsurface flow to sustain the needs of industries, power plants and cities. Therefore for support units requiring yields of more than 10,000 m³/day, it becomes necessary to release water from reservoirs and canals (Karanth, 1997).

In the desert terrains of Rajasthan, where the water table has commonly sunk to depths more than 120 m, ample supply of water has been recovered by tapping the sand-covered channels of streams and rivers related to the drainage system of the River Saraswati, that disappeared a few thousand years ago (Figure 5.8). The discovery of these water-bearing buried channels is attributed to the interpretation of satellite images. There are also pockets of water, in curvilinear chains, in the heart of the Thar desert (Rajawat et al, 1999). These freshwater-bearing horizons occur at depths of 60–250 m and 30–50 m (Soni et al, 1999). Significantly, despite intensive use, these reserves of freshwater have not depleted, nor is there any decline in the discharge of wells.

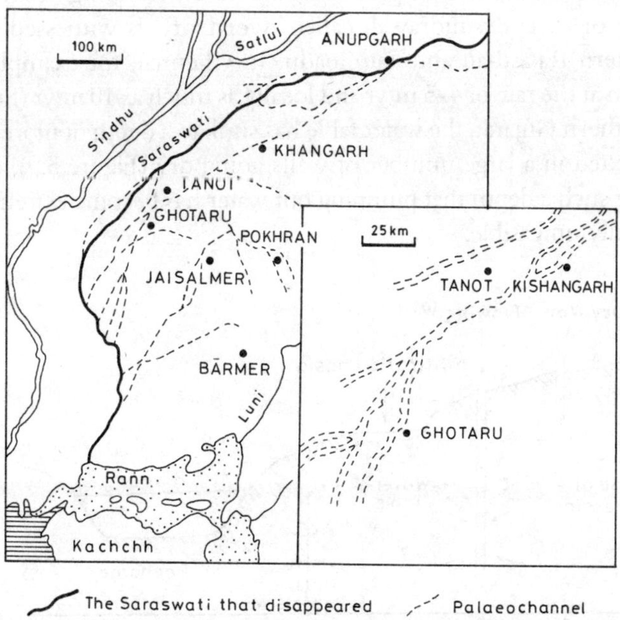


**Figure 5.8.** Abandoned channels related to the River Saraswati that disappeared a few thousand years ago, and are now buried under desert sands in the Jaisalmer–Hanumangarh region, have been found to yield an uninterrupted supply of freshwater in ample quantities (Based on Rajawat et al, 1999).

**Table 5.4.** Replenishable groundwater resources in India (After Subramanian, 2001)

| State | Resource (million ha-m/yr) | State | Resource (million ha-m/yr) |
|---|---|---|---|
| Andhra Pradesh | 4.34 | Maharashtra | 3.88 |
| Arunachal Pradesh | 0.14 | Manipur | 0.012 |
| Assam | 2.35 | Meghalaya | 0.043 |
| Bihar & Jharkhand | 3.38 | Mizoram & Nagaland | 0.005 |
| Delhi | 0.05 | Orissa | 2.33 |
| Goa | 0.061 | Panjab | 1.8 |
| Gujarat | 2.26 | Pondicherry | 0.017 |
| Haryana | 0.85 | Rajasthan | 1.62 |
| Himachal Pradesh | 0.036 | Tamilnadu | 3.02 |
| Jammu & Kashmir | 0.44 | Tripura | 0.06 |
| Karnataka | 1.62 | Uttar Pradesh & Uttaranchal | 8.05 |
| Kerala | 0.81 | West Bengal | 2.07 |
| Madhya Pradesh & Chhattisgarh | 5.97 | | |

## LOWERING OF THE WATER TABLE: A SCOURGE

Obviously, the withdrawal of groundwater causes depletion of groundwater reserves, manifest in the lowering of the water table. Pumping out of groundwater creates a **cone of depression** around the well (Figure 5.9). This happens if the rate of inflow is less than the quantity of withdrawal. The larger the number of wells, and the faster the rate of withdrawal, the greater would be the lowering of the water table. This practice of excess withdrawal, called **overdraft**, is witnessed in northern Gujarat, western Rajasthan and Tamilnadu. In Chennai, for example, the water table has fallen at the rate of 4–5 m/yr and locally as much as 10 m/yr (Subramanian, 2001). In northern Gujarat, the water table has sunk to a depth of more than 100 m. This has resulted in a large number of wells going dry (Figure 5.9), or the water level reaching such a depth that pumping out water has become extremely difficult or economically unfeasible.

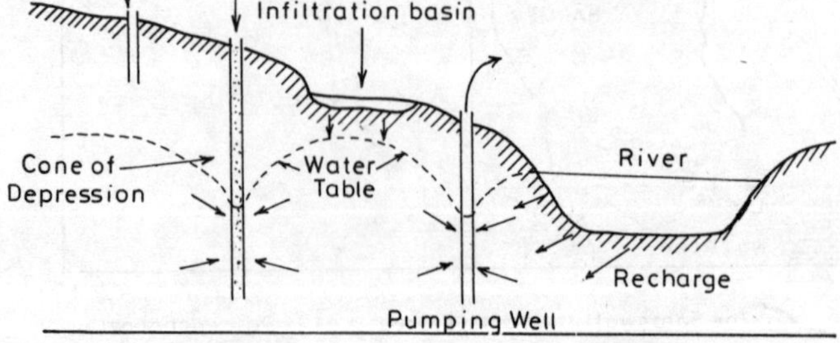


**Figure 5.9.** Heavy pumping out of groundwater (withdrawal) creates cones of depression in the water table around the wells. Coalesence of cones from a number of wells lowers the water table. As a result, shallow wells become dry and springs lose their discharge.

The consequences of the lowering of the water table are very grave. Shallow wells and springs become dry. In Panjab, over-exploitation of groundwater caused the water table to go down by 3–4 m in the last two decades, at the rate of 20 cm/yr (Tarlochan Singh, 1995). In Tamilnadu, over-pumping has lowered the water table by 25–30 m. Many parts of Chennai now face this problem.

Another consequence of the lowering of the water table in areas close to the coast is the **incursion of saltwater** from the sea (Figure 5.10). Saline water seeps in to fill in the empty space in the depleted aquifer. This renders the coastal groundwater unfit for consumption. In parts of Tamilnadu, saltwater invaded aquifers at the rate of 120–150 m/yr. In the Vasai–Viran belt of the Maharashtra coast (north of Mumbai), saltwater from tidal inlets has ruined the waters of wells (Figure 5.11). Against the recharge of 720 ha-m of rainwater, the withdrawal in the period 1991–92 was 765 ha-m, and the well density increased from 34 wells/km$^2$ to 45 wells/km$^2$ in the period 1996–97 (Duraiswami et al, 2000). The entire narrow coastal plain of Saurashtra is affected by the intrusion of saline water. In the 160-km Una–Madhavpur stretch, the salt-affected area dramatically increased from 35,000 ha in 1971 to 100,000 ha in 1977; and the salinity of 12,500 wells (including those 2.5–4 km inland from the sea) increased from 500 mg/l in 1970 to 2,500–7,500 mg/l in 1977, thus affecting 280,000 people and rendering unproductive nearly 100,000 ha of cultivable land (Ghassemi et al, 1995). It is evident that the overdraft has proved to be not only counter-productive in inland regions, but is also an irredeemable scourge in the coastal belt. The problem can be solved by preventing over-withdrawal or by building freshwater storage.

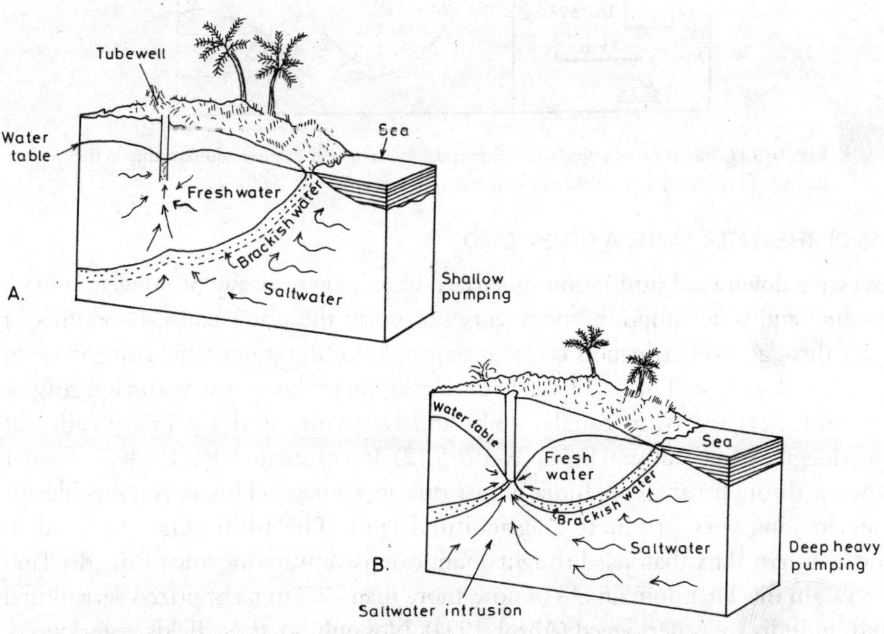


**Figure 5.10.** Intensive withdrawal of groundwater causing depletion of freshwater in aquifers along the coast lets in salt water from the sea.

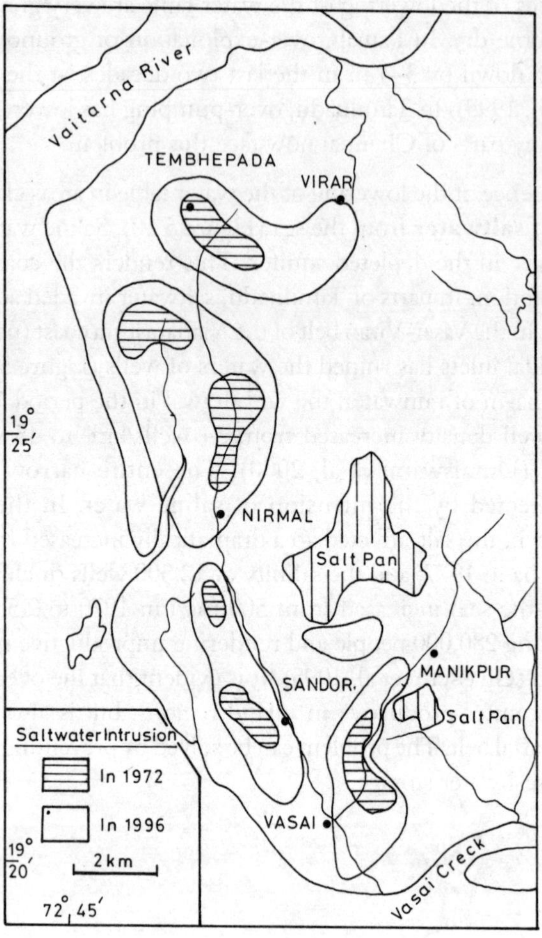


**Figure 5.11.** Incursion of saline water from tidal inlets into freshwater aquifers, due to the increase in the number of wells per unit area (Duraiswami et al, 2000).

## RISE OF THE WATER TABLE: A GEOHAZARD

Excessive downward infiltration of surface water due to heavy prolonged rainfall, flooding and water impoundment causes a rise in the water table. The influx of water through over-irrigation is also responsible for the water table rising close to the ground surface. This is a hazardous development, known as **waterlogging** of the land. Seepage from canals, tanks and reservoirs is the primary cause of waterlogging of agricultural fields (Figure 5.12). It is estimated that 15–40% of water flowing through canals in India is lost due to seepage. This is responsible for waterlogging very productive agricultural land. The Indira Gandhi canal in northwestern Rajasthan has brought about extensive waterlogging of the dry land (~ 5%) in the Thar domain. As of now, more than 12.7 m ha of prized agricultural lands in India are waterlogged (Abrol, 1994). Not only are these fields waterlogged, but more than 11 m ha have fallen victim to the scourge of salinization and alkalinization.

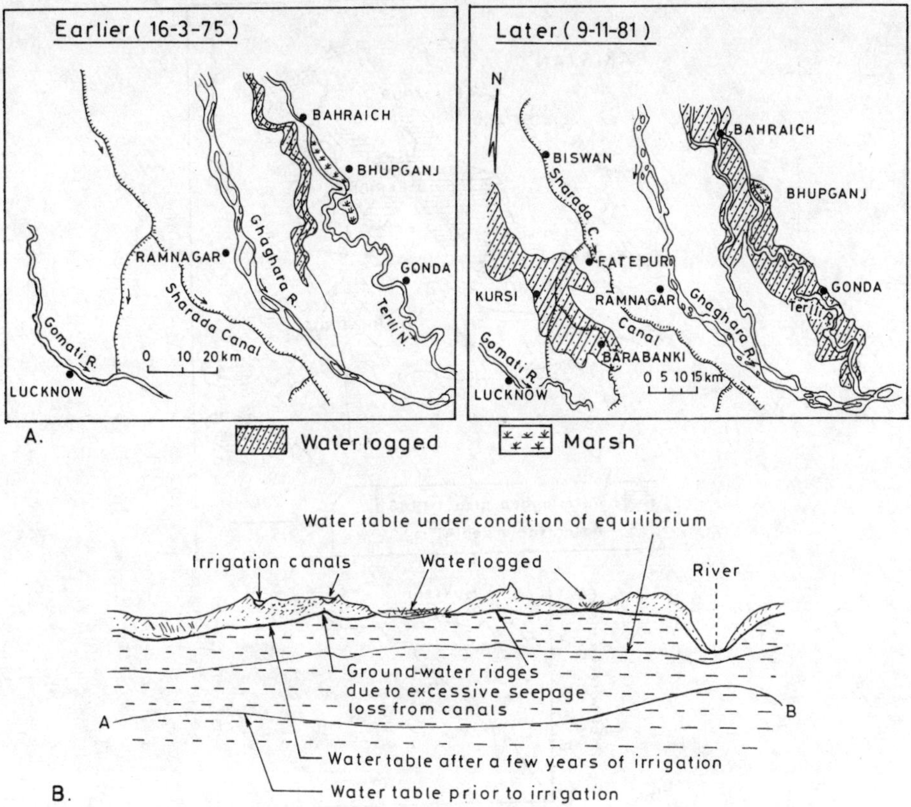


**Figure 5.12.** (A) Seepage from the Sharada canal increased the waterlogged area in irrigated parts of northeastern Uttar Pradesh in the period 1975–81 (Chaturvedi et al, 1983)
(B) Diagram shows how the rise of water table causes waterlogging in the canal-irrigated area.

In waterlogged soils, water rises to the root zones of plants, driving out air from the interstitial space in the zone of aeration. This results in the cessation of all processes that need oxygen. Plant growth is consequently retarded. Moreover, mosses and aquatic weeds get the upper hand, placing crop plants in a disadvantageous position. Consequently, the productivity of these plants falls considerably—by as much as 300%.

In parts of southwestern Panjab (Figure 5.13), where there is intensive irrigation, the water table rose at the rate of 20–50 cm/yr in the period 1975–95, resulting in the enlargement of waterlogged areas from 60,090 ha in 1978 to 104,000 ha in 1997 (K.P. Singh, 2000).

Waterlogged ground loses shear strength—the ability to bear loads of buildings and other utility structures. This is manifest in the subsidence of foundations and the cracking of walls. Also, the basements of buildings may fall victim to flooding, the walls may be affected by severe dampness, and there may be chemical damage to concrete structures due to chlorite and sulphate in the water.

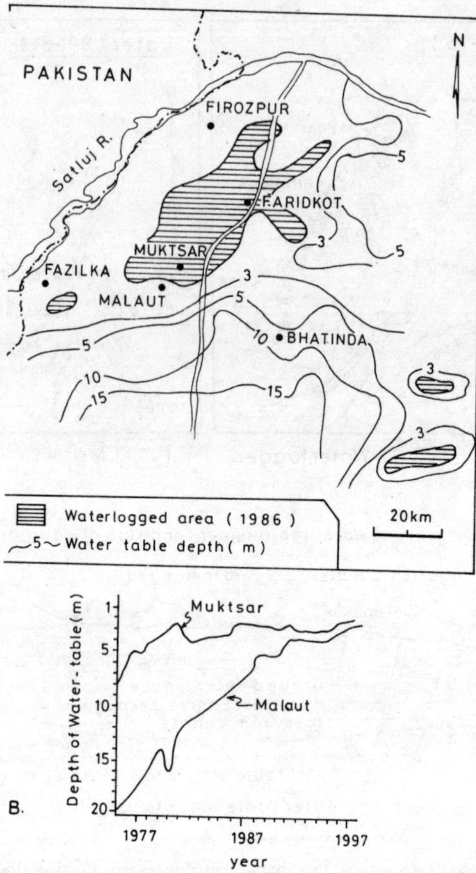


**Figure 5.13.** (A) Waterlogged area in southwestern Panjab where farmers indulge in intensive irrigation. This waterlogged area in 1986 became considerably larger by 1997.

(B) The water table in Muktsar and Malaut rose rapidly and came too close to the ground surface (Singh, 2000).

## Groundwater Pollution: Natural Causes

Groundwater, lying in aquifers, may get contaminated due to natural causes. This is attributed to certain minerals and metals that form the constituents of these aquifers. For example, the minerals fluorite ($CaF_2$) and fluorapatite ($Ca_5PO_4F_3$) occurring in granites, pegmatites and dolomites contribute to the fluorine content in groundwater in many parts of Karnataka, Tamilnadu, Andhra Pradesh, Gujarat, Rajasthan, Madhya Pradesh, Uttar Pradesh and Panjab. For some underprivileged people in the coastal regions of Kerala, the high incidence of a heart-muscle disease is attributed to high levels of thorium and cerium—and low amounts of magnesium—in the thorium-rich sands (Valiathan, 1998). Arsenic poisoning in groundwater has assumed alarming proportions in the floodplains of West Bengal and Bangladesh—with poisonous water emerging from 50–80 m deep aquifers. This poisoning has affected a sizeable number of people very badly. One view is that arsenic is released from the sediments by acid formation, in which *Thiobacillus* (a bacteria) plays a very active role (Khan et al, 2003). Another view is that this arsenic mobilization is associated with a recent inflow of carbon. This is indicated by a concentration of radiocarbon-young methane that results from pumping groundwater for irrigation purposes, to a level where dissolved arsenic is at a maximum (Harvey et al, 2002). In the Dhaka area of Bangladesh, as a result of excessive withdrawal of groundwater, the water table has declined rapidly—upto 20 m in the last six years (Pinaki Roy, 2003). In Bangladesh, as a whole, the water table has dipped at the rate of 2 m/yr. The subject is dealt with in detail in Chapter 19.

Groundwater is also subject to indirect natural pollution, which can be human-induced. Not only do rocks/sediments in watershed areas and stream basins affect groundwater pollution, but so does the highly polluted riverwater itself (Figure 5.14). This is evident from the sediments of the Ganga between Kanpur and Unnao in U.P., which show very high levels of toxic elements—an increase in upto 23.6 for chromium, 14.7 for cadmium, 12.2 for tin, 3.2 for zinc, 2.6 for copper and 1.6 for nickel (Ansari et al, 1999). The groundwater in Chennai has been affected by the polluted Kuvam river. This is reflected in abnormal trends in health among people who live along the riverside, due to the presence of trace elements of chromium, cadmium, zinc and copper (Selvaraj, 2000).

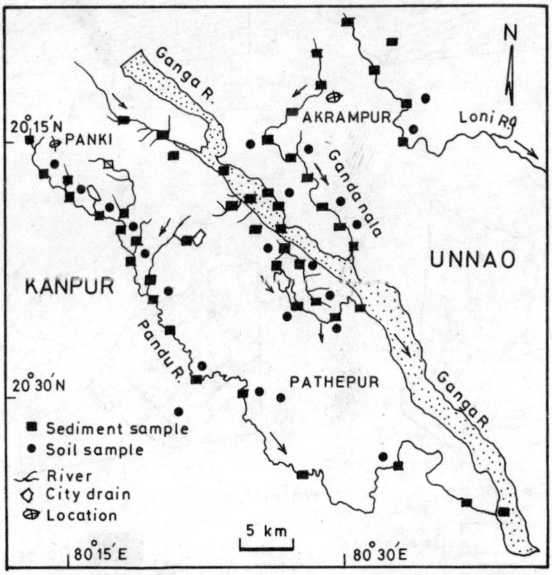


**Figure 5.14.** In the alluvial Kanpur–Unnao tract, the metal pollutants carried by the Ganga have found their way into the groundwater, making well water unfit for consumption (Ansari et al, 1999).

Chemical knowledge of soils, sediments and rocks helps geologists to identify the provenance (source) of contaminating minerals/elements, and to trace the paths along which the toxic material was carried to reach the places where they occur today. Geologists must also suggest ways to check the ingress of contaminated water into the source area and also to deviate the path of the flow. Thus, the role of geologists is becoming increasingly crucial as more and more groundwater bodies become polluted and poisonous.

## Mountain Springs and their Dwindling Discharges

In hills and mountains where the water table meets or intersects the surface, underground water comes out in the form of **springs** and oozes up as **seepages** (Figure 5.15). In the plains, swamps and marshes develop where the water table rises above the floor of topographic depressions. Groundwater comes out at a contact point between permeable and impermeable layers (Figure 5.15A), or where a dyke or reef cuts the permeable layer (Figure 5.15B). It may move out along a fracture or fault (Figure 5.15D), or come out through networks of cavities and caverns in areas rich in limestone-dolomite (Figure 5.15D). The yield of springs depends not only on the rate of flow of the groundwater, but also on the rate of infiltration through soils and rocks.

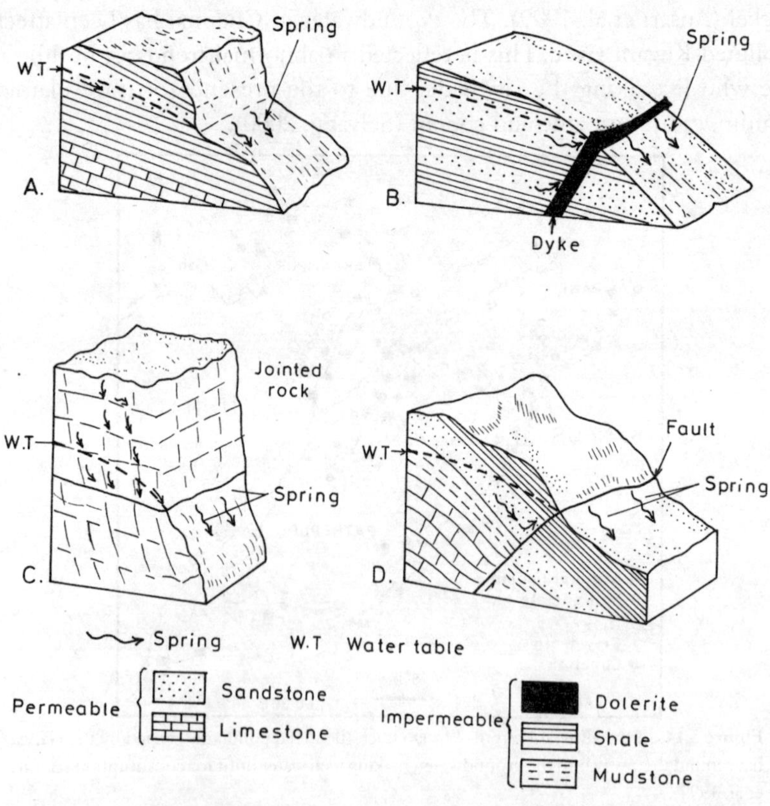


**Figure 5.15.** Types of springs showing how springs originate in different geological settings:

(A) Along the contact between permeable and impermeable layers.

(B) Where a dyke cuts the water-bearing layer.

(C) When a fault intervenes and provides a passage.

(D) In jointed or cavernous limestone terrain.

The **recharge of springs** depends on: (i) the slope of the terrain, (ii) the texture of the soils or structural discontinuities in rocks, (iii) the nature and extent of the vegetation that covers the recharge area, and (iv) the intensity and duration of rainfall. Gentler slopes promote larger volumes of water percolating downwards; so do porous and permeable soils, and fractured, sheared, jointed and crushed rocks. Vegetation characterized by prolific foliage is very conducive to the recharge groundwater.

Extensive deforestation, the subsequent denudation of the land, and also intensive excavation on hill slopes for road construction and mining, results in considerable reduction in spring discharge. This is manifest in springs becoming seasonal—spouting water only during the rainy season, and often becoming wholly dry. In the catchment of the Gaula rivulet in southcentral Kumaun, where the watershed lost 13.1% of its forests in just 22 years, the decline in water varied between 25% to 75% in 5–50 years (Bartarya and Valdiya, 1989). In cultivated and moderately grazed land in the Dugar valley in Pauri Garhwal, the spring discharge decreased by 36.4% in the period 1994–98 (Negi et al, 1998).

Significantly, in the land covered by pine forests in the same region, the reduction in spring discharge was 100%, implying that pine forests are more vulnerable to water deficiency.

### THE IMPLICATIONS OF REDUCED SPRING DISCHARGES

Springs and water seepage feed large streams and rivers. Weak spring discharge means a reduction in the flow of rivers during the dry season. The Gaula rivulet in southcentral Kumaun, for example, registered 29.2% decline in its discharge in the period 1951–70 and 38.5% between 1971–81 (Bartarya and Valdiya, 1989). This directly affects rivers in the Himalaya, which will carry considerably less water in the time to come. Already the ratio of outflow between the rainy and dry seasons is 120:1 for the Karnali, 150:1 for the Sapta Kosi, and 180:1 for the Kali Gandaki (Uprety, 1988). Significantly, the dry-season flow of water in the deforested Lesser Himalayan rivers of Nepal is 60,000 to 30,000 times less than the flow in the rainy season (Kattleman, 1987). A similar situation exists elsewhere in the Lesser Himalaya. This too-much–too-little water syndrome is an indicator of the onset of desert conditions in the Lesser Himalaya (Valdiya, 1987; 1997b). As a matter of fact, the appearance of drought-tolerant xerophytes on the slopes of the deforested Lesser Himalaya indicates that Nature is already preparing to cope with the problem of aridity.

## A Blueprint for Water Management

Water management implies environmental reconstruction in headwater areas (Haigh, 2000b). The integrated action plan for management of water resources entails maintaining and preserving water, land and vegetation in catchment areas. This endeavour also links water management to social and economic development. Water management, in a restricted sense, implies ensuring a continual and adequate supply of requisite quantity of water, but without endangering the life of the resource. This involves four principal programmes:

(i) The use of the finite resource has to be controlled rigorously through a regulated system of distribution, on the basis of genuine needs, both in terms of quantity and quality.

(ii) The aquifer must be simultaneously recharged with appropriate methods—watershed treatment, development of spring sanctuaries and construction of percolation tanks and pits.

(iii) Water harvesting, wherever possible, must be pursued, as a programme of top priority.

(iv) Diversion of surplus water from regions of abundance to areas of deficiency, by diverting flows through canals, must be undertaken as a national plan. This requires building dams and embankments to raise water levels. In other words, more reservoirs are required in river basins.

## RECHARGING AQUIFERS

The key to successful groundwater management lies in understanding the processes of groundwater recharge, and the success of this venture depends on high-quality mapping of the ground with its soils and rocks (Summers, 1998). This mapping should not be restricted only to delineating groundwater sources, but also to charting zones of protection from geohazards, like pollution. Recharge can be accomplished by inducing infiltration of rainwater, and water in channels, rivers, lakes and pits. By flooding fields (Figure 5.16A) during rains and when the rivers are in spate, considerable amounts of groundwater recharge can be done. The impounded and stored water will eventually percolate down and fill the aquifers. Canal irrigation also accomplishes much (Figure 5.16B). Likewise, floodwaters can be allowed to fill depressions, or specially excavated trenches, tanks and furrows lining highways and railway lines. In urban centres, groundwater recharge takes place from leaking water mains, septic tanks, leaching sewers, and over-irrigated gardens and parks.

One of the most effective methods of aquifer recharge is the injection (pumping in) of water through a series of **injection wells** dug into floodways (Figure 5.16 left *inset*). In the Kasuliwara area in the Saraswati valley of north Gujarat, excessive withdrawal of groundwater has lowered the water table to great depths. Injection wells more than 120 m deep have shown that partially empty aquifers are easily filled by percolating water (Phadtare and Bradley, 1983). Five 40–60 m deep wells, dug 130 m apart along the Narwana canal in Haryana raised the groundwater level appreciably—1.5 m in the first three weeks, and 0.32 m in the next two weeks, the rate of recharge being 10–12 litres per second per delivery siphon (Gupta, 1975). The **percolation tanks** in Maharashtra have proved very effective in recharging groundwater in the basaltic lava terrain—especially in the districts of Aurangabad, Osmanabad, Beed, Ahmadnagar, Solapur and Sangli (Karmarkar et al, 1995).

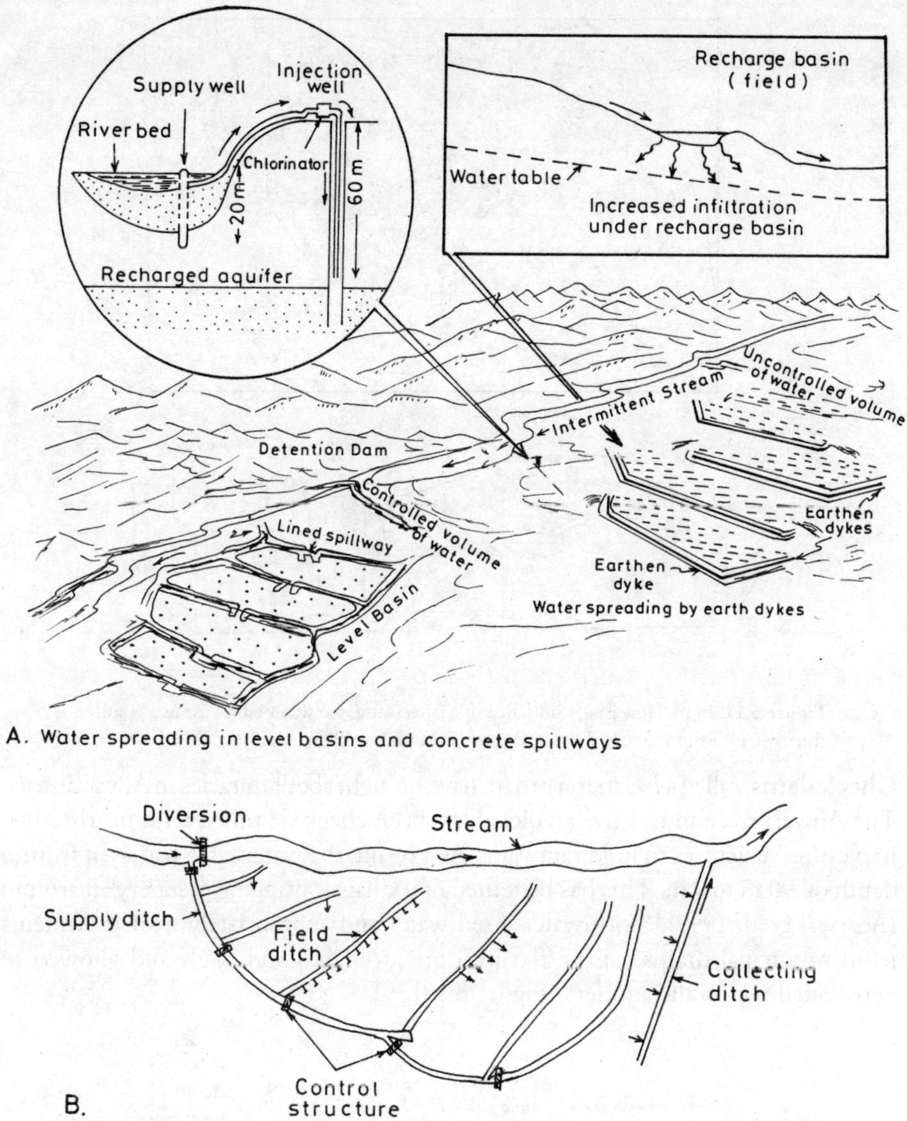


**Figure 5.16.** (A) Groundwater is recharged by spreading floodwaters on fields through well-laid canals and channels and letting water percolate down. Inset to the left shows pumping out of water from a riverbed and injecting it through the injection well to recharge groundwater. Inset to the right demonstrates the percolation of water stored or held in a tank/ditch till it reaches the aquifer.

(B) A network of canals that is allowed to leak water all through is quite effective in recharging groundwater.

The chronically drought-prone Ralegaon–Sidri area (Figure 5.17) in Ahmadnagar district was brought under the watershed development programme and it recorded a considerable rise in the water table in all the wells (Deshmukh, 1995). This resulted in foodgrain production rising from 29 tonnes/year in 1976 to 138.6 tonnes/year in 1991.

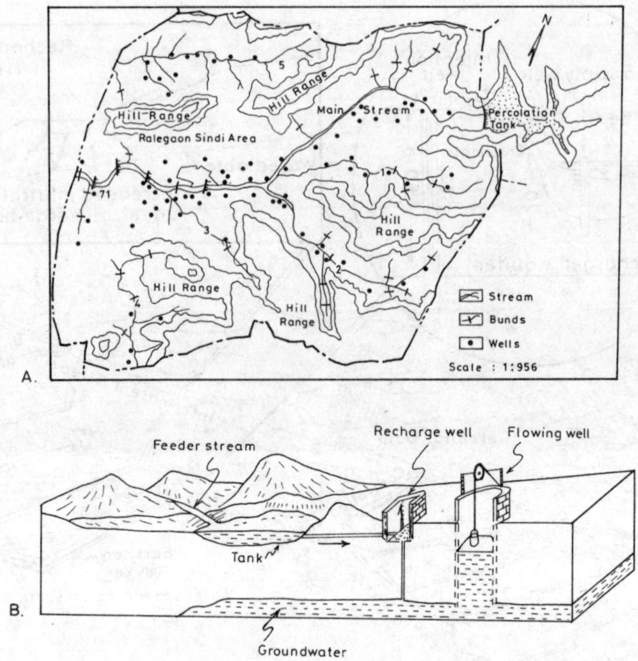


**Figure 5.17.** In the Ralegaon–Sidri area in Maharashtra, the water table rose as a result of intensive effort in watershed management (Deshmukh, 1995).

**Check dams**, called *johad* in Rajasthan, have brought about miracles in Alwar district. The Anwari river and its five rivulets have been check-dammed with nearly 3,000 harvesting structures to hold rainwater. As a result, the water table has risen from a depth of 90 m to 9 m. This has benefited 650 villages, bringing greenery all around (Agarwal et al, 1999). Stormwater, used water and appropriately treated effluents from municipal drains can be fed into pits, trenches and wells and allowed to percolate down to the aquifers (Figure 5.18).

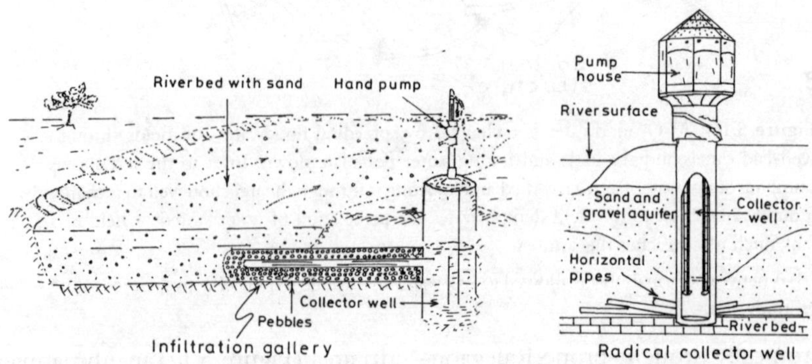


**Figure 5.18.** (A) Schematic diagram showing how stormwater from a drain or stream can be injected —after proper treatment—into an aquifer to recharge it. The system also includes a recovery unit.

(B) Reclaimed water is injected into a brackish-water aquifer in order to improve the quality of water (Based on Dillon, 2002).

**Table 5.5.** Groundwater recharge through percolation tanks, water harvesting structures and soil conservation measures (Sikka, 2002).

| Region | Measures | Rise in Water Table (in metres) |
|---|---|---|
| Central Maharashtra | Percolation tanks | 2.5 |
| Coimbatore district, T.N. | Percolation ponds | 0.8–2 |
| Chinnatekur, Kurnool, A.P. | Check dams, bunds and weirs | 0.5–1 |
| Anantpur district, A.P. | Water harvesting structures | 2–3 |
| G.R. Halli, Karnataka | Water harvesting structures | 1.5 |
| Parbani, Maharashtra | Nala, bund, gully plugs | 0.3–2.5 |

## SPRING SANCTUARIES

In hilly and mountainous terrain, the most effective and appropriate method of augmenting spring discharge is the development of **spring sanctuaries** (Rai et al, 1998). These subcatchment-scale watersheds are enclosed by constructing stone walls or fences to ward off grazing animals, poachers, and fuel–fodder gatherers. This measure allows the unimpeded growth of natural vegetation. Additionally, leafy trees and shrubs that generate litter, grasses and water-absorbent mosses are grown and nurtured. All efforts are made to promote the unhindered growth of multi-cultural vegetation within the fenced area (Figure 5.19A).

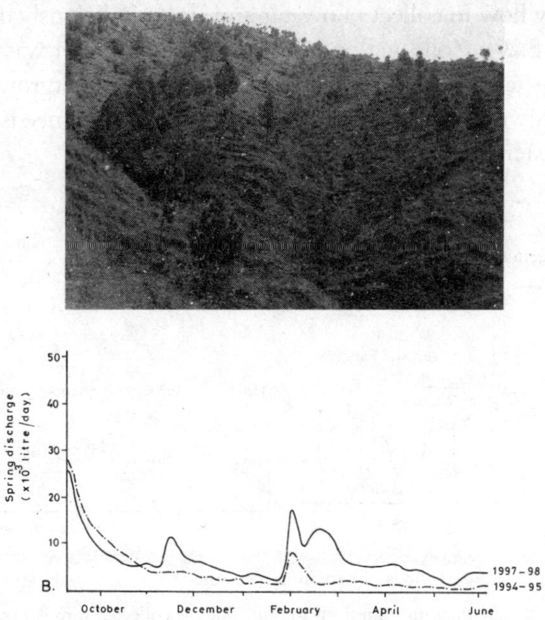


**Figure 5.19.** (A) In order to develop a spring sanctuary, a subcatchment is chosen and fenced or stone-walled in order to allow free growth of vegetation. Recharge through percolation of rainwater is accomplished by cutting trenches, ditches and pits in areas selected on the basis of favourable geology.

(B) Spring discharge perceptibly increased in the year 1998 compared to that in 1994–95 in a spring sanctuary in Pauri Garhwal (Negi et al, 1998).

The next step is the construction of **percolation tanks, trenches, ditches** and **pits** to induce the infiltration of runoff rainwater. These are dug along or across identified shear zones, faults, fractures and joints, or porous-permeable horizons of rocks. If necessary, tanks or trenches are filled with very coarse, permeable gravel in order to protect them from choking by garbage or debris. This will also effectively minimize evaporation loss. It is also necessary to construct check dams (Figure 4.11) in gullies—not only to control erosion, but also to store water that would eventually percolate down to reach the groundwater. An attempt to develop a spring sanctuary in the subcatchment of the Dugargad stream in Pauri Garhwal was well rewarded. The spring discharge (Figure 5.19B) increased almost fourfold in just four years: from 595 l/d in June 1995 to 2,170 l/d in June 1998 (Negi et al, 1998).

In a novel experiment in a barren watershed in Chittoor district (A.P.), a stream was impounded by erecting a dam. The reservoir water percolated underground and filled a well dug immediately downstream of the dam (Muralidharan, 2003). This well water is now transferred 1.5 km downstream, through pipes, to the wells of the village Jhillayagiripalli, where the people now draw water at the rate of 20 l/min. The village wells being 25 m below the feeder well, the water flows with the help of gravity and without electricity.

## HARVESTING WATER

People of old knew how to collect rainwater and use it judiciously during periods of scarcity (Figure 5.20). Collected on hill slopes, the rainwater was fed into large wells (*bhandaras*) or tanks, and then transferred through underground tunnels to wells or *bawdis* in towns. This practice of rainwater harvesting needs to be revived on a much more extensive scale all over the country.

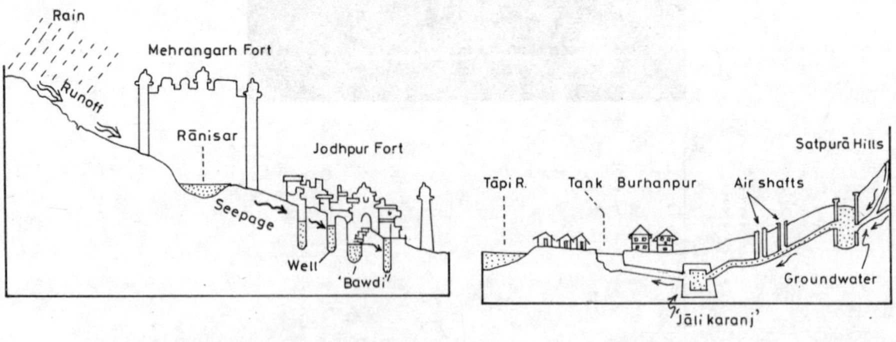


**Figure 5.20.** Time-tested methods of harvesting rainwater.

(A) At Jodhpur in Rajasthan the runoff on the hill slope is collected through converging channels into a large tank within the fort, and then led through subterranean conduits to the wells in the city located at the foot of the hill.

(B) In Madhya Pradesh the runoff is collected in large deep wells and then tunnelled down to the township of Burhanpur, on the banks of the Tapi river (Modified after Agarwal et al, 1999)

In Maharashtra, the Marathas (in the time of Shivaji) dug cisterns in volcanic ash beds between lava beds on the hilltop on Sinhagarh Fort, and stored enough water to meet the requirements of a large garrison for the whole year. Residents of Israel, a desert country, surround their underground cisterns with boulders to protect them from heat storms. The sides and bottoms of the cisterns are lined with plastic or polythene sheets to prevent seepage. The surface of the tankwater is covered with monomolecular films of hexadecadol, which prevents evaporation of water despite letting in air.

It may be mentioned that bentonite clay is a good lining material for storage tanks and ditches. When wet, the clay swells fifteen times its volume, and thus works as a sealant, effectively controlling seepage.

In all places where the annual rainfall is 50 mm or more, water harvesting can be practiced (Figure 5.21). The cumulative harvesting would amount to 90 mm more than all the surface water considered usable today (Kayastha, 2000). It is thus the most effective and reliable solution to the problem of water shortage. Rainwater harvesting is needed to supplement moisture in rainfed areas. Its contribution to increase in agricultural productivity is 1–1.5 tonnes per hectare (C.D. Thatte, per. com. 2002).

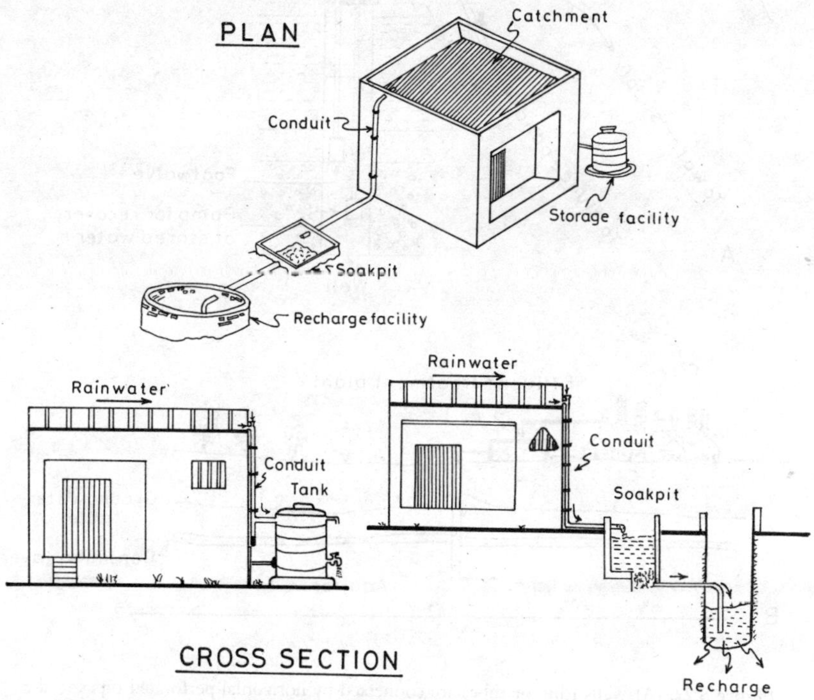


**Figure 5.21.** Every drop of water needs to be saved. Water collected on rooftops is fed into wells—either to recharge groundwater or to be stored. It can be filtered through a pit to remove pollutants.

Another aspect of water harvesting is to catch water from seasonal riverbeds (without surface flow during dry seasons). The water-saturated sand of the riverbed can be tapped by constructing infiltration galleries and radial collector wells (Figure 5.22). The wells are connected to the river bed by horizontal perforated pipes.

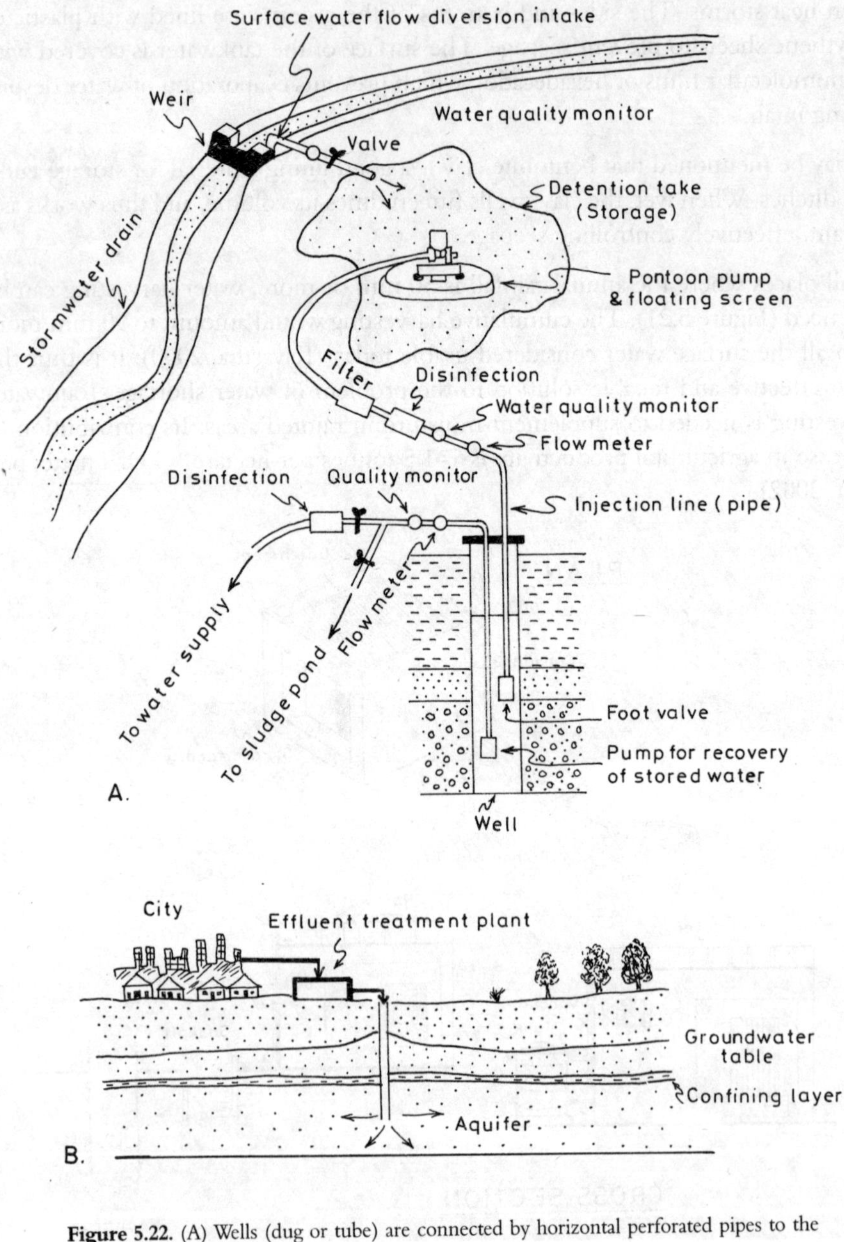


**Figure 5.22.** (A) Wells (dug or tube) are connected by horizontal perforated pipes to the water-saturated sand of a seasonal stream. Around the pipe, pebble pitching is done to prevent sediments from entering the pipe (*Catchwater*, 2002, 4(1): 9–10).

(B) A tank or lake can, likewise, be connected to the recharge well in order to raise the water table in the dug well.

### CURBING OVER-CONSUMPTION OF WATER

Appreciating the fact that reserves of good quality water are very limited, and that it costs much to treat polluted water, it is imperative that this precious asset is distributed among users on the basis of genuine needs. The needs must be determined in terms of both quantity and quality. Industries and agriculture should not be allocated the same water that is used for drinking and bathing. There is thus a strong need for developing **two systems of water supply**—one for drinking and bathing and the other for watering gardens, washing clothes, utensils, etc and for industries, power plants, etc. This would require laying parallel systems of water distribution through two different pipelines. Even if this is expensive and labour-intensive, this measure would have to be introduced to save a precious resource that has become extremely scarce. Also, being an earth resource, water must be priced appropriately.

In the matter of irrigation, farmers must change the prevailing practice—irrigation should be carried out only for limited durations and using limited quantity of water. Resorting to **drip irrigation**—which has brought about miracles in Israel—not only leads to a 50–70% saving of water, but also reduces growth of weeds, lowers fertilizer consumption, accelerates crop maturing and ensures 20–25% higher yield.

Our reserves of water are very limited, but the demands are growing exponentially. It is therefore necessary that every drop of water is used judiciously and with utmost care. One alternative is that water supply be rationed and priced, in other words, marketed—allowing the market to determine the price of water. This would also encourage efficiency in the water supply system.

Groundwater should be utilized, and not mined. 'We must make the right choices, and our actions must be guided by more than technology and economics. The fact is that water is essential to life. This lends an ethical dimension to every decision we make about how it is used, managed and distributed. All living things must get enough water before some get more than enough.' (S.L. Kayastha, per. com., 2002).

# 6

# Regulating Mining Activities

## Non-renewable Mineral Resources

Society would not be what it is today if mineral resources had not been discovered. Minerals are inseparably intertwined with the industrial, commercial and economic structures that humans have built for their survival and development.

उदक्रंद दविणोदा बाज्यर्वाकः
सलोकं सुकृतं पृथिव्याम्

Sacred earth has been explored to give wealth to mankind.

— *Yajurveda*

In one form or another, minerals are used to generate energy, fertilize agricultural fields, produce defence materials, etc, all of which play a major role in the sustenance of modern civilization. Modern development activities are dependent on the availability of a large array of minerals—coal, petroleum, natural gas, ores of iron, aluminium, copper, lead, zinc, uranium, and fertilizers like phosphate and potash. However, these coveted mineral resources—which predominantly occur in the earth's subsurface or lie covered under thick layers of sediment or soil—happen to be *non-renewable* assets. Most of these were formed long, long ago, through several geological time spans, and undergoing processes that took millions of years! The processes occurred in very specific geological locations and under specific physico-chemical conditions. The process of the formation of mineral deposits was extremely slow. Therefore, there never can be **second crops** of minerals in contemporary times. Once the deposits are used, they are gone for ever—**wasting assets** that can neither be renewed nor replenished.

### GROWING PRESSURE ON WASTING ASSETS

Wasting assets are under tremendous pressure, generated by ever-increasing demands. This is evident from the fact that compared to the mining of 20 minerals valued at Rs. 700 million in 1950, India today produces 84 minerals valued at Rs. 450,000 million (Agarwal, 2001). The demand for various minerals is more than doubling every nine to fifteen years. This naturally calls for sustainable resource management by the government, or in other words a **Mineral Resource**

**Management Policy**. This policy should embody the following elements: (i) appraisal of the extent and potential of existing reserves, (ii) planning their exploration and extraction in a manner that would ensure their availability for future generations, (iii) exploring new deposits in newer domains, and (iv) developing mining and metallurgical techniques that will cause minimal or no waste, no pollution and no ultimate degradation of the land and the environment.

Exploration and mining of mineral deposits requires an understanding of the nature and origin of the host rocks and the processes that bring about the transformation in their character, increase in their qualitative value and the eventual concentration of the ore-forming or mineral-forming elements into mineral deposits. For this, geologists must know the mode of occurrence and location of such deposits.

In the process of extracting mineral deposits, land is excavated and defaced, hydrological systems are disrupted or impaired, and wastes are generated in very large quantities. Some toxic elements in abandoned mine material or mine waste may become exposed to exogenic processes, resulting in an increase in their toxic content. Geologists are required to tackle these problems and suggest measures for the restoration of the environment. Mineral deposits, when wisely exploited, have the potential of producing real wealth.

## The Status of India's Mineral Reserves

Going by the production figures in recent years, India is wholly or largely self-sufficient in 35 minerals (Tables 6.1 and 6.2). These minerals constitute the raw materials of basic industries and include iron ores, aluminium ores, raw materials for cement, fertilizers, glass, ceramics, paints, and a large number of inorganic chemicals. In the matter of iron ores, manganese, barytes and talc-steatite, India is a producer of global importance. The production of bauxite, coal, chromite and magnesite are also quite significant in the global perspective (Table 6.2). However, reserves of crude petroleum and natural gas, and non-ferrous minerals like copper, zinc and lead are not sufficient to meet the demands of Indian industries.

India is woefully deficient in cryolite, spinning-grade asbestos, borax, acidic filters, industrial diamonds and alloy metals.

The last three decades have witnessed a fifty-fold increase in production of minerals. The life indices of several minerals have also increased. India now ranks among the top ten countries in the world, with respect to producing minerals such as mica, barytes, metal-grade chromite and bauxite, thermal-grade coal and lignite, and manganese ores.

Besides mineral deposits, there are about 300 **hot springs** in ten **geothermal fields**, whose heat-flow values vary between 83–650 mW/m$^2$, with thermal gradients between 70–100°C/km (Shanker, 1988). The important geothermal fields are in Puga (Ladakh), Tattapani (M.P.), Jalgaon (Maharashtra), and Manikaran (H.P.) and are significant sources of underground energy . Water from many of these springs also possesses therapeutic value.

**Table 6.1.** Status of minerals in India (M.H.R. Rao, 2001).

| Grouping | Abundant | Adequate | Deficit | Scarce |
|---|---|---|---|---|
| Fuel minerals | coal | | coal (cooking), petroleum, crude oil | |
| Ferrous minerals | iron ore | chromite (metallurgical grade) | chromite (refractory grade), manganese ore | nickel, tungsten, cobalt, molybdenum, vanadium |
| Non-ferrous minerals | bauxite (metallurgical grade), copper, lead | zinc | bauxite (chemical refractory grade) | antimony, gold, platinum group of minerals, silver, tin |
| Industrial minerals | barytes, ball clay, bentonite, calcite, dolomite (excluding low silica grade), felspar, fire clay, Fuller's earth, garnet, gypsum, kaolin, limestone, magnesite, mica, pyrite, ochre, quartz and silica sand, quartzite, sillimanite, zircon, sheelite | corundum, graphite, rock salt, vermiculite, wollastonite | apatite, rock-phosphate, asbestos, fluorite, kyanite | andulusite, borax, diatomite, native sulphur, potash |
| Precious stones | | | | diamond, emerald, sapphire, ruby |
| Other minerals | granite (dimension stone), marble | | | |

**Table 6.2.** Reserves of mineral deposit in India (Based on Anon, *Indian Mineral Year Book*, 1998–99; Dasgupta, 1999; Varma, 1999).

| Mineral | Recoverable Reserve (in million tonnes) | Life (in years) |
|---|---|---|
| Mica | 60 | 300 |
| Barytes | 87.1 | |
| Chromite | 86.2 | |
| Chrome ores | 146 | 93 |
| Diamond | 981,515 | |
| Talc-Steatite-Pyrophyllite | 214.7 | |
| Bauxite | 2,462.4 | 299 |
| Manganese ores | 167.3 | 65 |
| Magnesite | 245.1 | 267 |
| Rock salt | 3.5 | |
| Sillimanite | 51.6 | 940 |
| Kyanite | 2.8 | 23 |
| Asbestos | 9,382.8 | |
| Coal | 206,239 | 410 |
| Lignite | 5,864 | |
| Lead ores | 179.1 | 67 |
| Zinc ores | 10.1 | |
| Gold ores | 67.9 | |
| Gold metal | 57.0 | 97 |
| Iron ores | 13,460 | 257 |
| Copper ores | 416.8 | 34 |
| Limestones | 75,678.9 | 367 |
| Dolomite | 4,386.8 | 357 |
| Rock phosphate | 145 | 130 |
| Tungsten ores | 54 | |
| Tin ores | 29 | 4 |
| Fluorite | 2.9 | |
| Nickel | 281 | |
| Potash | 404 | |
| Molybdenum ores | 2.5 | |
| Felspars | 31,286 | |
| Fireclay | 518.1 | |
| Silicon minerals | 2,402.1 | |
| Gypsum | 237.6 | |
| Ilmenite | 90.1 | |
| Rutile | 6.4 | |
| Petroleum oil | 764.6 | 15 @ 35 m t/yr |
| Petroleum gas | 636 billion $m^3$ | 24 @ 35 m t/yr |

## The Conservation of Mineral Resources

Keeping in mind the finite nature of mineral deposits, it is important that these resources are conserved and that extraction methods remain rigorously scientific. Scientific methods should follow these considerations:

(i) During extraction (mining and metallurgical) there should be minimum waste.

(ii) A judicious blend should be maintained between high-grade ores and sub-marginal or low-grade ores.

(iii) Low-grade ores should be subjected to the processes of **beneficiation**, if amenable.

(iv) The ecological health of terrains subjected to mining activities must be maintained at all costs.

Since technological advances in extraction and processing of ores have made it possible to profitably extract even low-grade ores, more and more low-grade ores are being treated as source material and not discarded as waste. Some examples are the ores of iron, manganese, lead, zinc, copper, gold and chromium.

**Table 6.3.** Estimated production of important minerals in India (Anon, *Statistical Outline of India, 2000-2001*)

| Mineral | Production (2000-2001) |
|---|---|
| Bauxite | 7,061,000 tonnes |
| Chromite | 2,007,000 tonnes |
| Coal | 323,000,000 tonnes |
| Copper concentrates | 149,000 tonnes |
| Gold | 4,053 kg |
| Gypsum | 2,951,000 tonnes |
| Iron ores | 77,700,000 tonnes |
| Lead concentrates | 58,000 tonnes |
| Limestone | 127,000,000 tonnes |
| Manganese ores | 1,580,000 tonnes |
| Mica | 1,100 tonnes |
| Natural gas | 28.4 billion $m^3$ |
| | 1.6 billion $m^3$ flared |
| Petroleum oil | 32,000,000 tonnes |
| Zinc concentrates | 359,000 tonnes |

## The Adverse Effects of Mining

As of January 1995, there were 9,000 mining leases for major mining ventures, with a cumulative lease area of more than 700,000 ha. There were also 3,500 large and small mines, including 580 mines of coal and lignite (Anon, 2000b). More than 800,000 ha of land in India are affected directly by quarrying and mining activities, and much larger areas are disturbed due to the activities associated with mining (Figure 6.1).

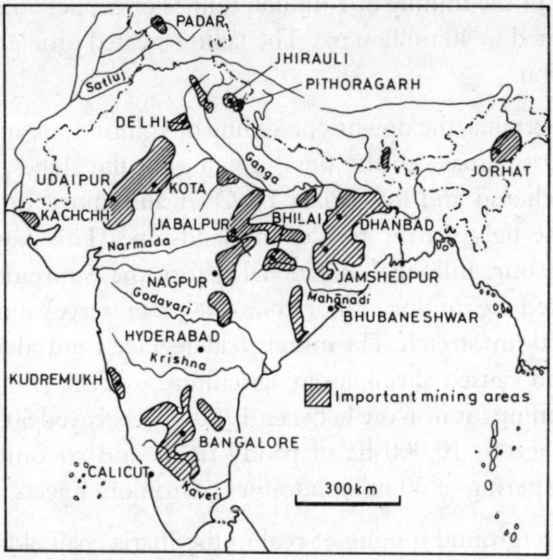


**Figure 6.1.** Extent of land affected adversely by various mining activities (After G.K. Datt in Valdiya, 1987).

Quarrying on a large scale has brought about profound landscape changes and caused serious land degradation. **Open-cast mining** alone accounts for more than 30% of the areas affected. In the processes related to mining, not only is vegetation destroyed extensively, but precious soil is lost forever in most cases.

Mining operations entail reshaping of the region's topography and the resultant generation of great volumes of debris, often incurring the disruption of both surface and groundwater systems (Plate 7). The network of roads to serve large mining operations aggravate the problem of land degradation. In the Gandhmadan Reserve Forest (Orissa), the mining of bauxite has endangered 2,700 species of flowering plants and 150 species of medicinal plants and herbs (Anon, 2000b). Sandstone quarrying in the Bijolia belt in Rajasthan—which saw a 36.5-time increase in the extent of excavation in just two decades—resulted in 47% decrease in forest cover, 14% decline in agricultural area and 67.5% increase in wasteland (Anon, 2000b).

The magnitude of the backlash of mining can be appreciated with examples from coal mining. The average depth of the **overburden** in open-cast mines is assumed to be about 100 m. A 100 km$^2$ coal field with an annual production of 200 million tonnes, would have generated 8,800 million m$^3$ of debris, at the rate of 4 m$^3$ for every tonne of coal produced. By AD 2000, the 270 million tonnes of coal produced from the Jharia coalfields are said to have resulted in degradation of land at the rate of 60 km$^2$/yr, and generated 700 million m$^3$ of wastes (Rai, 2000). Consequently, 49,000 ha of land will have to be abandoned, with 11,000 ha having been covered by wastes and 9,916 ha sinking due to underground mining. In the Sigrauli coalfields in southwestern U.P., for the 2,700 million tonnes of coal extracted, more than 7,800 million m$^3$ of debris had to be disposed of in 6,560 ha of forested area. In the brief period of 1988–89 in Jharia, mining operations decimated 36,385 ha of forest land (Rai, 2000). In the mining of 7 million tonnes of copper annually, the tailings generated amounted to 40 million m³. The tailing created problems like landslides and water pollution.

In mountainous terrains, the downslope sliding of voluminous amounts of mining wastes causes great damage to the flora growing on the slopes, and springs and streams get smothered and lost (Plate 7). Over and above this environmental destruction, is the ugly scarification of the landscape. This was the case in the picturesque Mussouri hills in Uttaranchal before the Supreme Court's orders drastically restricted the mining of limestone (400 m t reserve) and phosphorite (45 m t reserve) to a 30-km stretch. The mining had degraded and adversely affected 36 km² of forests and caused shrinking in agriculture by 55% (Roy, 1996). In Goa, indiscriminate mining for iron ore between 1961–83 destroyed 50,000 ha of forests, and caused damage to 10,000 ha of paddy fields and coconut groves due to indiscriminate scattering of 30 million tonnes of iron ore rejects.

Likewise, the underground mining of coal in the Jharia coalfields (which entailed the withdrawal of coal from the 80–200 m deep horizon) resulted in subsidence of the ground by 6–25 cm by the end of 1960 (Figure 6.2), leading to the deflection of sewers, streams and even the Barakar river. The resultant floods have understandably been disastrous in the coal fields. By AD 2000, about 10,000 ha of land had sunk in the Jharia coalfield area (Rai, 2000).

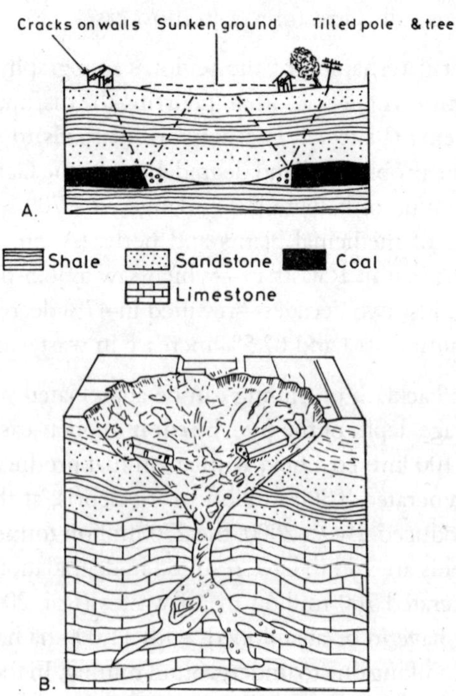


**Figure 6.2.** Underground mining causes subsidence of the ground, with serious consequences.

Underground fire in coalmines (Table 6.4) is another hazard that is difficult to control. The fires in the Jharia coalmines have destroyed tens of millions of tonnes of coal and threatened the safety of the towns above ground. The burning of coal releases gases like sulphur dioxide, carbon dioxide, carbon monoxide, methane, etc as well as coal dust—all of which are hazardous to life.

Deeper excavation in both surface and underground mining causes the water table to sink, drastically in some places. The consequence is the drying up of springs and wells in neighbouring areas. Exposure of mined rocks to the atmosphere initiates weathering, with the inevitable formation of sulphates which are highly soluble water-pollutants. The Tiri stream, emanating from the Zawar mines in Rajasthan, is polluted with cadmium (0.6–1.6 ppm), zinc (>50–<50 ppm) and lead (0.5–1 ppm) for more than 15 km downstream (Ahmad et al, 1996). This is responsible for the elimination of, among others, the plant *Wringtonia*. In the Zawar mines, the groundwater contained dissolved arsenic between 1.4 ppb–13.8 ppb, while the groundwater at the Khetri copper mines contained arsenic below 1 ppb (Madhavan and Subramanian, 2000).

**Table 6.4.** Mine disasters in India during the period 1908–95. (After Parasuram and Unnikrishnan, 1998)

| Date | Mine | Deaths | Cause |
|---|---|---|---|
| 1908 | Nadu Khas | 20 | Explosion |
| 1910 | Dishergarh | 11 | Roof-fall |
| | Sitalpur | 14 | |
| | Nambarg | 14 | |
| 1936 | Poldih | 205 | |
| 1953 | Majri | 11 | |
| 1956 | Burra | 28 | |
| 1958 | Chunalani | 175 | |
| | Baro Dhemo | 29 | |
| 1965 | Dhori | 260 | |
| | Chinakuri | 200 | |
| 1973, March | Noonidih | 48 | Explosion |
| 1975, August | Kessurgarh | 11 | Roof-fall |
| 1975, November | Sillewera | 10 | Inundation |
| 1975, December | Chasnala | 375 | Inundation |
| 1976, September | Central Saunda | 10 | Inundation |
| 1976, October | Sudamdih Shaft | 43 | Explosion |
| 1979, January | Baragolai | 16 | Explosion |
| 1981, June | Jagannath | 10 | Fire |
| 1982, July | Topa | 16 | Roof-fall |
| 1983, September | Hurriladih | 19 | Inundation |
| 1993, November | Tipong | 9 | Explosion |
| 1994, January | New Kenda | 55 | Fire |
| 1995, September | Geslitand | 64 | Inundation |

## Ecofriendly Methods of Mining

All mining ventures must be preceded by an **Environmental Impact Assessment** (EIA). EIA involves appraising the adverse effects of a venture on the resources of the projected area—water, soil, forests, pastures, agricultural fields, etc—and the cost estimate of restoring the ecological health of the area in case of pollution.

These ecofriendly measures include:

(i) Systematic and appropriate methods of waste-disposal and regeneration of vegetation in the mined area.

(ii) Safe stacking of the top soil and ensuring its use it after mining, to cover the excavated part, so as to enable the reappearance of vegetation (Figure 6.3).

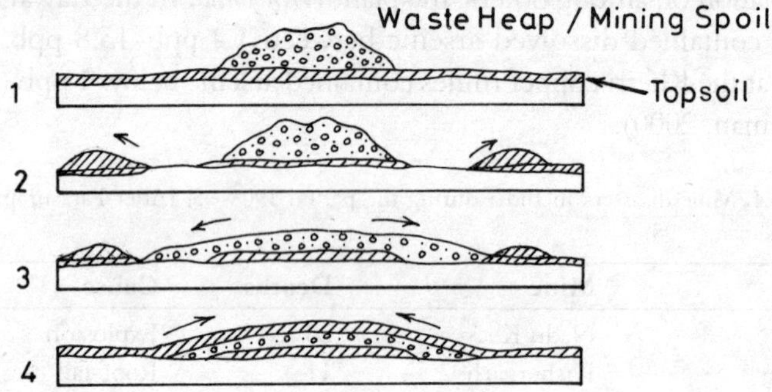


**Figure 6.3.** Scraped from the ground, the topsoil is stacked away safely before the start of mining. The stacked material is later spread over the mined part or mine wastes so that vegetation reappears after the end of mining operations (After Bradshaw and Chadwick, 1980; Bradshaw, 1997).

(iii) Prevention of slope failure in the mined tract during and after the mining operations by maintaining the stability of hill slopes, as well as of the walls of the cuts.

(iv) Avoidance of mining in ecologically sensitive and geodynamically active areas/belts where earthquakes, mass movements, etc are recurrent. If this is not possible, effective preventive measures are to be carried out simultaneously.

### RESTORING MINED LANDS

For adequate recovery of mined lands, the topsoil of the area that is to be mined is stacked away safely, instead of being disposed. The overburden from the contiguous cut is then dumped in the trench formed previously, as a result of removal of ores (Figure 6.4). This process is continued until the last trench is excavated. The dumped material is then levelled and covered by the topsoil that was earlier stacked away. Finally, appropriate species of grasses, shrubs and trees are planted to form a protective cover on the dumped mass. Liberal use of fertilizers and municipal sludge ensure quicker growth of vegetation.

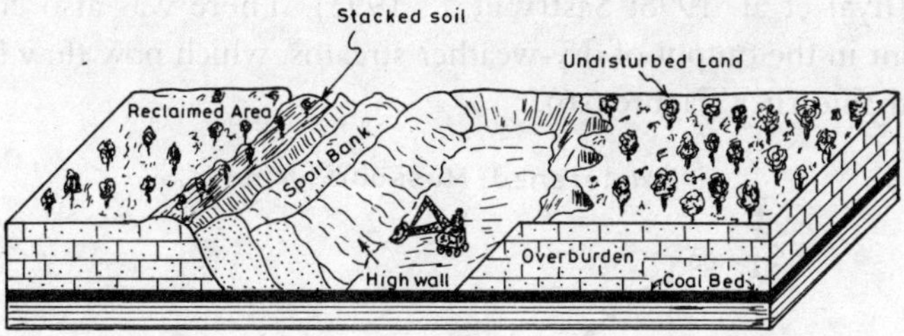


**Figure 6.4.** Mechanized mining has given rise to a big pit. The excavated overburden material is dumped successively in the contiguous, earlier-cut portion. The dumped material is levelled, and finally covered with the topsoil that was stacked away elsewhere.

## STABILIZING DEBRIS DUMPS

The dumps of debris and **spoil banks of tailings** are stabilized by diverting and draining out water from the mined area and then covering them with a carpet of vegetation. A network of drains (Figure 6.5) would ensure that no mischief-causing water remains within the mined area. Stability is achieved by leaving the outer part of the ore-body, along with its overburden, intact so that it serves as a barrier of sorts.

The debris is stacked on the uncut part and later covered by the topsoil, treated with organic fertilizers and seeded. If the mining waste contains a toxic element derived from the deposits, it will be necessary to neutralize it by application of the appropriate material. Lime counters the effects of acid-producing materials (mainly sulphur in sulphides), and sludge (domestic or municipal) or organic manure neutralizes alkaline wastes. The addition of chopped straw, wood pulp and shredded wood not only provides protection from the sun's rays, but also ensures retention of moisture in the seeded soil.

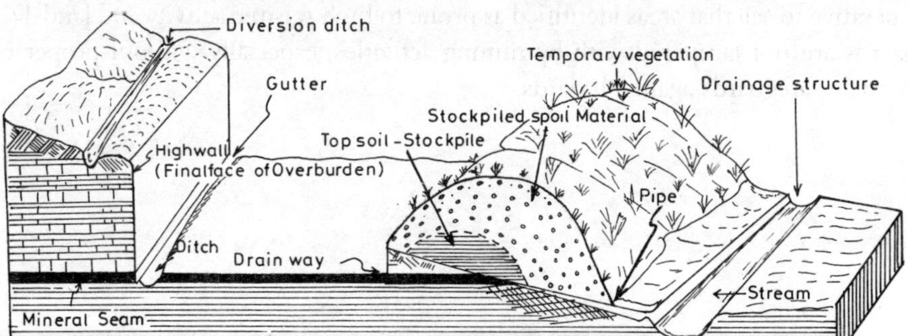


**Figure 6.5.** Construction of diversion drains along and over the wall of the mining cut, and across the stockpiles of spoil and wastes, contributes to the stability of debris dumps. The dumps are then seeded with grasses, shrubs and trees to stabilize the ground.

Biological and engineering remedial measures over a period of six years, in the limestone mines of the Mussouri hills, successfully ensured reduction in the rainwater runoff from 57% to 37% and the sediment outflow from 550 t/ha/yr to

8 t/ha/yr (Juyal et al, 1998; Sastry et al, 1995). There was also considerable improvement in the output of dry-weather streams, which now flow for 120–150 days more than earlier (Figure 6.6).

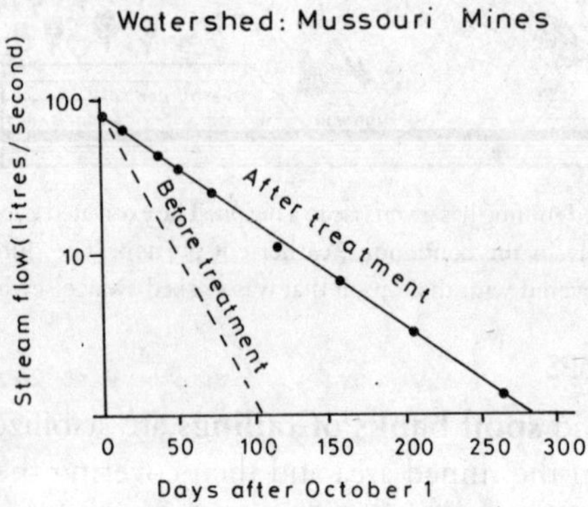


**Figure 6.6.** Dry-weather streams in the Mussouri hills showed considerable increase in water discharge after the mined area was treated with biological and engineering measures (Juyal et al, 1998).

Before embarking on mining ventures, governments and concerned functionaries must ensure that the lease applications for mining are supported by documents containing the following material: (i) **Environmental Impact Assessment**, (ii) a five-to-ten year plan for landscape restoration of mined area, and (iii) a working plan (Kilmartin and Haigh, 1998). These plans should encompass not only the assessment of the environmental conditions, but also the likely fallout in the entire area to be affected by the mining, including by the ancillary activities of mining. Unless there is a provision for lasting compensation to the people who are eventually affected by mining, the extraction of minerals should not be permitted. It is also imperative to see that areas identified as prone to high seismic activity and landslide hazards are not tampered with by mining activities, especially without proper or adequate safeguards against hazards.

A.

B.

Land is a cherished but finite asset:

(A) An agricultural field in the Karnataka plateau.

(B) Terraced crop field amidst a forest, in the Kosi valley in the Kumaun Himalaya, depicting the intimate relationship between forests and agriculture.
(Photos: K.S. Valdiya)

Plate 2

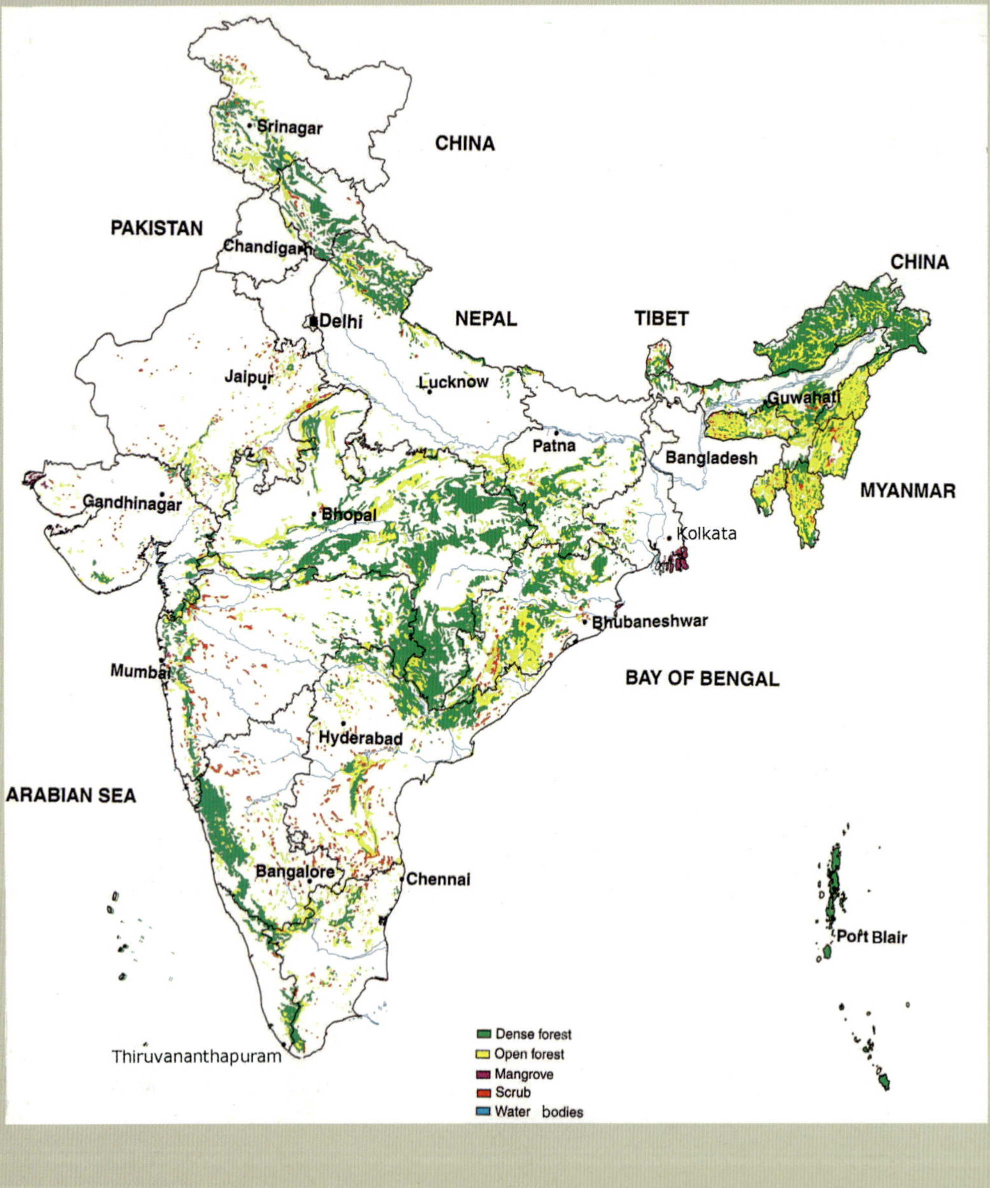


Forest cover in India in 1998. Only 19.39% of the land surface in India was under cover of forests in both good and bad conditions (*State of the Forest Report*, 1999). The parts that are not covered by forest—not shown as green in the map—are exposed to the relentless, unremitting actions of wind and water.

Badland or ravine land in the valley of the Mahi river in Gujarat.
(Photos: L.S. Chamyal)

Plate 6

(A) The Bhagirathi and the Alaknanda converge at Devprayag, Uttaranchal, to make the Ganga. (Photo: K.S. Valdiya)

(B) The Kaveri near Kollegal, Karnataka. (Photo: M.J. Haigh)

Plate 7

A limestone mine in the Mussouri hills, Uttaranchal:

(A) Debris generated due to downslope sliding of mining waste, killing vegetation.

(B) Hydrological system disrupted and agricultural fields damaged due to excavation for extracting minerals.
(Photos: S.K. Bartarya)

# Plate 8

(A) Formation of the ~90 km-long and 2.5–3.7-m high ridge, known as the Allah Band, due to the 1819 earthquake in northwestern Kachchh.
(Photo: C.P. Rajendran)

(B) Water spouted in a large number of places in the Banni plain and the Rann of Kachchh, when the region was struck by an earthquake in January 2001.
(Photo: C.P. Rajendran)

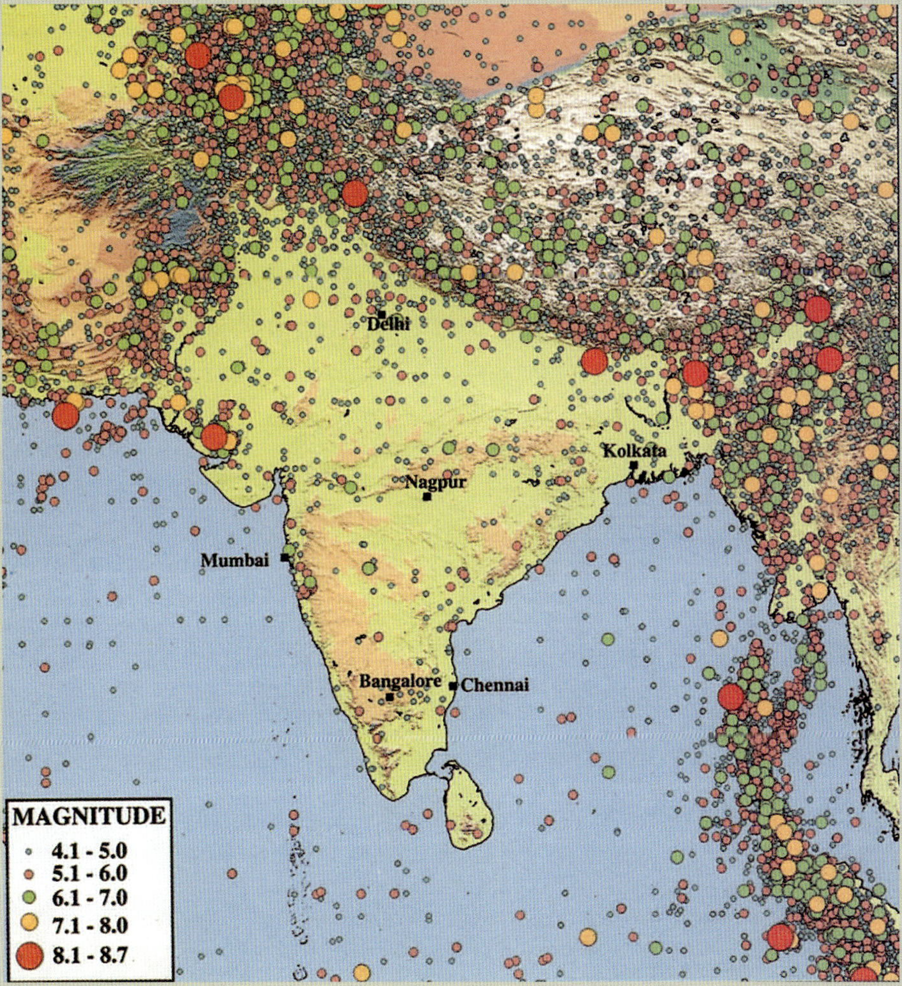


In northern India, the epicentres (black dots) are concentrated in the foothills of the Himalaya where the Peninsular Indian Plate is sliding under the Himalaya along a tectonic plane. In the southern Indian Shield, earthquakes (open circles) arise due to movements along older faults, shear zones and rift valleys that were formed in the distant geological past. (Cover of Annual Report of CMMACS, Bangalore, 2003. Courtesy: Gangan Prathap)

Plate 10

Landslides in the Main Central Thrust zone in the Guptkashi area (Uttaranchal) characterized by a multiplicity of faults and shattered rocks.
(Photos: M.P. Sah)

(A) The Bhagirathi branch of the Ganga was blocked for more than 14 hours in August 1978, due to the dumping of a mass of sediment placed by debris flow in the Kanodia stream.
(Photo: K.S. Valdiya)

(B) Gully erosion, of the kind seen near Nandidurga in Karnataka, provides large volumes of sediments.
(Photo: K.S. Valdiya)

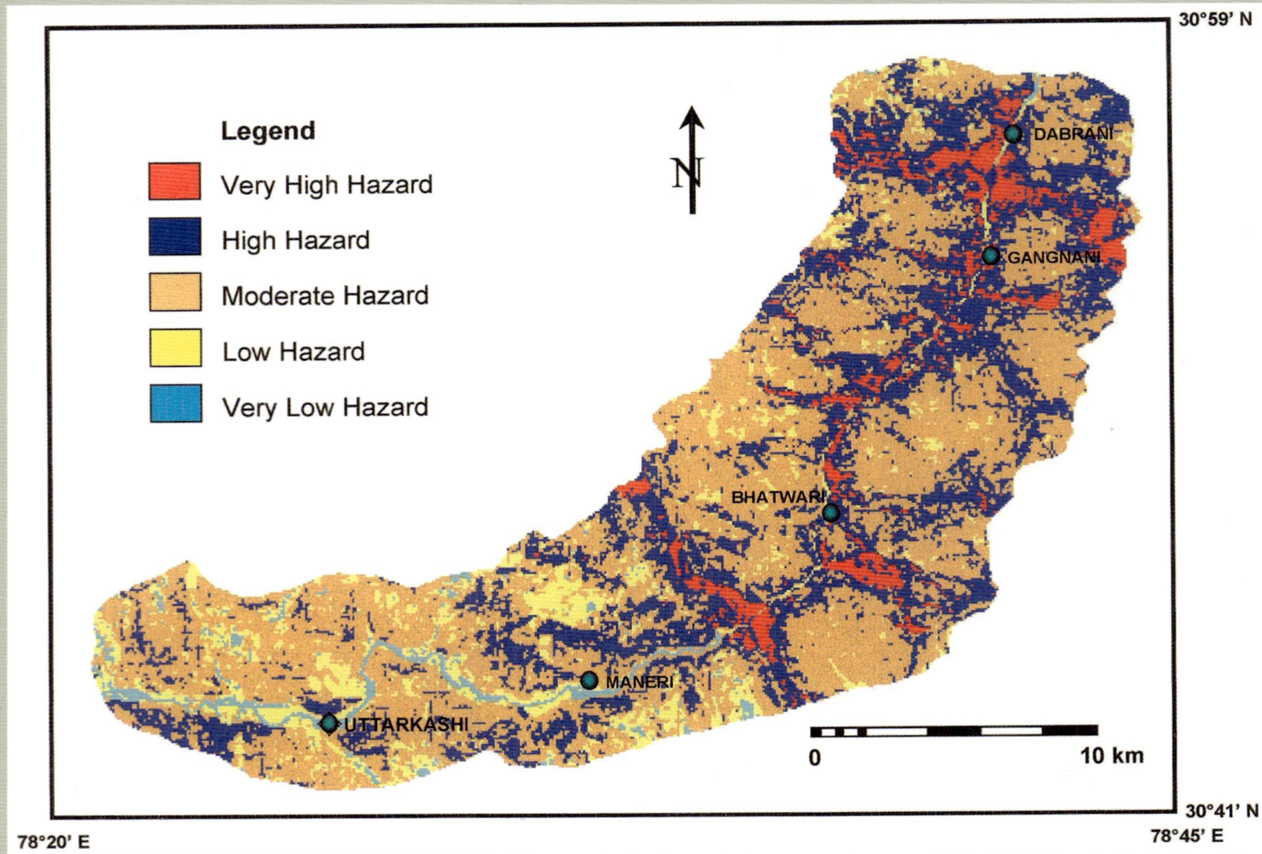


Landslide hazard-zonation map of the Bhagirathi valley of the Ganga (northeast of Uttarkashi), prepared from satellite imagery. The map shows zones of varied degrees of risks resulting from slope failure. (Gupta et al, 1999. Courtesy: R.P. Gupta, 2000)

(A) Raphstreng glacial lake in Bhutan; to the left is the Lugge Tso. The dam-burst occurred in October 1994.
(From Mool, 2001)

(B) Tso Rolpha glacial lake in Nepal.
(From Mool, 2001)

Plate 20

Growth of a delta in the Kulekhani reservoir in Nepal. The deposit comprises primarily sand, silt and gravel. The gravel occurs at the top.
(Galay et al, 1995).

# 7

# Coping with Natural Hazards: Providing a Framework of Strategy

## Swift, Sudden Phenomena

Sudden and swiftly occurring geological phenomena often destabilize the natural configurations of life and threaten the balance of ecosystems. These developments may cause harm to people, their homes and their institutions. Suddenness is not always the rule—slow-moving landslides, the slow rise of the water table and swelling of soil clays have caused harm to houses and civil structures.

Naturally occurring spasmodic phenomena which cause damage and are fraught with risks are called natural hazards. **Geohazards** include earthquakes, volcanoes, cyclones, tornadoes, storm surges, avalanches, landslides, ground subsidence and floods. The slow rise of the sea level is soon going to become a hazard for low-lying coastal towns.

Left to themselves, most sudden swift phenomena would be just geological events. Only when they affect people do they become hazards. Avalanches, for example, are quite a common geological occurrence in the Himalaya, but become a hazard in the pilgrim routes to Badarinath and Amarnath, or on the strategic road to Leh. The swift flow of water in floodways is, likewise, a normal occurrence. But when people encroach upon the floodways, the floods bring distress year after year in the Indo-Gangetic Plain.

Of the more than three million people killed all over the world in the years between 1960–90, as a consequence of various natural hazards, nearly 75 per cent belonged to underdeveloped and developing countries. In contrast, only 0.7 per cent of the fatalities occurred in developed countries. It is the poorest section of society—accounting for more than 80 per cent of the world's population—that suffers greatly from natural disasters. This is largely because this section cannot afford to move from the places where they have been living, even in vulnerable and hazard-prone areas. The number of people affected by various natural disasters is increasing at the rate of 6 per cent per year. This, contrary to one's expectations, does not mean that disasters are increasing at this rate, but that more and more people are now compelled to occupy the areas that are prone to natural disasters (Table 7.1).

**Table 7.1.** Human lives lost due to various natural disasters in India (Sharma et al, 2000).

| Year | Floods | Cyclones | Hailstorms | Earthquakes | Total |
|---|---|---|---|---|---|
| 1990–91 | 1,320 | 979 | - | - | 2,299 |
| 1991–92 | 1,185 | 304 | - | 768 | 2,257 |
| 1992–93 | 1,193 | 497 | - | - | 1,690 |
| 1993–94 | 1,690 | 318 | - | 7,930 | 9,946 |
| 1994–95 | 2,038 | 247 | 59 | - | 2,344 |
| 1995–96 | 2,072 | 361 | 31 | - | 2,464 |
| 1996–97 | 2,069 | 1,719 | 40 | - | 3,828 |
| 1997–98 | 1,560 | 216 | 247 | 39 | 2,062 |
| 1998–99 | 2,567 | 1,292 | - | 106 | 3,965 |

It would be pertinent to explain the meanings of the words hazard, disaster and risk here.

> A **hazard** is defined as a phenomenon that poses a threat to people, structures or economic assets. A **disaster** is serious disruption of the functioning of society, causing widespread human, material or environmental losses, which exceed the ability of the affected society to cope with, by using its own resources. A **risk** is a measure of the expected losses due to a hazard (Gupta et al, 2001).

### HUMAN BEINGS: BOTH VICTIMS AND PERPETRATORS

Not all disasters are natural. Some are brought about by people through acts of commission and omission. In the Western Ghat and the Himalaya, landslides have been triggered due to extensive deforestation and cutting of roads, or due to the occupation and cultivation of unstable hillslopes. Building obstructions within floodways—the normal course through which excess runoff flows during the rainy season—has brought deluge after deluge in the Indo-Gangetic Plain.

Humans have courted risks by living under the shadow of crumbling mountains, and invited disasters by building houses in old, abandoned channels of rivers that have been changing courses through time. By removing the protective shield of plants, they have induced not only deep erosion, but also landslides in the hills. By building dams across rivers, coastal erosion has been aggravated, because the sediments that normally build or nourish beaches are held back in artificial reservoirs. People have been forced to pay heavily for what they have done to Nature.

## Facing the Challenge of Disasters

The occurrence of hazards cannot be prevented in all cases, but their consequences can be made less severe, less risky. This can be achieved by taking timely action in coping with hazards. These measures reduce the intensity and extent of the disasters, by changing the trends of the happenings and by distributing the distressful, harmful effects.

The nature of geohazards and the geodynamic conditions in the Indian subcontinent make it unrealistic to hope to stop disaster-causing geological phenomena. People have to learn to live with natural hazards in certain areas and prepare to cope with all its adverse effects. Communities vulnerable to hazards must be resilient enough to withstand the stresses of hazardous events and keep on functioning. This implies that all life support systems be located—or relocated—in places that are out of harm's way, and that essential facilities are designed to continue functioning during landslides, floods, storms, earthquakes, etc. This calls for a shift in strategy from **resistance** to **resilience**.

In order to make communities capable of meeting the challenges of hazards, it would be necessary: (i) to keep the hills under the protective cover of vegetation, so that enough rainwater percolates underground, and thus minimize the chances of floods in the floodplains, and also stabilize slopes against failures, (ii) allow the floodplains to generate wetlands to accommodate floodwaters, (iii) permit beach nourishment through formation of sand dunes, etc. These measures imply a policy of **anticipation** rather than **reaction**. Simultaneously, it is imperative that all buildings and infrastructure are strengthened by means of: (i) building codes, (ii) engineering designs, and (iii) construction practices.

There is, theoretically, another option: evacuating the vulnerable or threatened area permanently or temporarily. In India, the only option is to leave temporarily—there is so little land left for resettlement. As a matter of fact, it is the pressure of rapidly growing population that is responsible for the occupation and use of hazard-prone lands. In the mountains, people are constrained to move up the steep, forested slopes; and in the floodplains the only habitable places left are floodways and abandoned channels. The concentration of people in cities, likewise, exposes them to the hazards of pollution.

For people living under such trying circumstances, the only option left is to undertake **hazard-preparedness programmes** in identified hazard-prone zones. To put it differently, it is imperative to make plans for coping with disasters, based on geological-geomorphological appraisals—the physical conditions, magnitude and patterns of past happenings, and the nature and degree of geophysical and landscape changes taking place in the present.

## Identifying Hazard-prone Zones

The primary prerequisite for hazard-coping measures is to prepare Hazard-zoning Maps. This exercise is based on knowledge of the geological setting, tectonic condition and topography of the land.

### MAPPING HAZARD ZONES

The preparation of hazard-zoning maps entails:

(i) Delineation of areas in accordance with the degree of hazard anticipated, based on an integrated assessment of the interplay of various causative factors. For instance, in the case of landslide hazard-mapping, considerations of geology, hydrology, landform, land use, and land management, as well as understanding triggering factors like earthquakes, cloudbursts, etc assume importance.

(ii) Delineation of hazardous areas based on facts observed during previous episodes. Once again, in the case of landslide hazard-mapping, such a map is expected to show both active and dormant landslides. The identification of such landslides can be done by interpreting satellite imagery and aerial photos taken during ground surveys, and by past experiences and the available body of information.

(iii) Assessment of the degree of probability of hazards in the identified areas and an anticipation of the consequences.

Hazard maps have to be developed by continuous matching of hazards inferred from a multitude of causative factors, and the actual hazardous events mapped through field surveys, till the best fit is found between the two. The ideal map is one which goes a little further, by accounting for problems in the making for which irrefutable evidence can already be seen on the ground. For instance, in landslide hazard-mapping, it would be important to recognize the landslides triggered by seismic activation of old landslides, or by the cascading effect caused by the possible formation and breach of a landslide dam (R.K. Bhandari, per. com., 2002).

Hazard maps, by themselves, are of little use unless they are made to a scale appropriate for practical application. Small-scale maps, as found in the *Vulnerability Atlas of India*, do help in macro-level planning, but one would require mapping on a scale of 1:25,000 or 1:10,000 for development planning. These utility maps must also be user-friendly. The present trend is to work towards evolving multi-hazard maps so that the consequent projection of risk scenarios are closest to ground reality (R.K. Bhandari, per. com., 2002).

Attempts to predict hazardous happenings are based on:

(i) Various signs of physical changes taking place in the natural systems. This is done: (a) through interpretation of aerial photos and satellite imagery, (b) field surveys and visual observations of common hazard indicators, (c) field instrumentation and monitoring, and (d) clues provided by global positioning system measurements.

(ii) Interpretation of the erratic behaviour of animals, birds and insects that have an extra-sensory perception of impending disasters.

(iii) Accumulated experiences of generations of people who have experienced the stress of disasters.

The Building Materials Technology Promotion Council, of the Ministry of Urban Development, has published the *Disaster Vulnerability Atlas of India* on a scale of 1.2.5 million (see Gupta et al, 2001 for some of these maps). According to the *Atlas*, the Indian subcontinent is among the world's most disaster-prone areas—nearly 54% of the land is vulnerable to earthquakes, 8% to cyclones and 5% to floods. Of this, 12% of the earthquake-prone land is liable to severe earthquakes (intensity

MSK[1] IX or more); 18% land is liable to MSK VIII (similar to the earthquakes at Latur and Uttarkashi); and 25% land is liable to MSK VII (similar to the Jabalpur earthquake). The *Atlas* presents state-wise hazard maps, and includes the union territories. Each administrative district boundary is clearly marked with its hazard intensity.

Remote sensing through satellite images (Plate 15) provides vital information—comprehensive synoptic multi-temporal coverage of large areas in real-time and at frequent intervals (Rao, 2000). An example is the INSAT picture taken on 9 May 1990, that helped in the timely evacuation of 170,000 people from the Andhra coast before it was ravaged by a cyclone. Likewise, the 29 October 1999 supercyclone that devastated a large part of the Orissa coast was predicted well in time through satellite pictures. The post-disaster imagery was also of great help during relief and rehabilitation measures.

## RISK ASSESSMENT

Risk assessment requires the preparation of a map of infrastructure which shows: (i) human settlements, including building topology, (ii) important buildings, roads, bridges, etc, and (iii) demographic maps showing distribution and density of population. This makes it easy to work out the elements at risk, the specific and total risks involved, and eventually estimate the extent of economic loss if the hazard strikes along the anticipated lines. Disaster scenarios, in any given situation, are usually developed as a strategy so that **Disaster Mitigation and Management Plans**, as well as the attendant contingency plans, could be made realistic.

## HAZARD PREPAREDNESS PLAN

The hazard preparedness plan embodies these agenda:

(i) Hazard-zoning maps pre-suppose the probability of an event occurring, at a given location and in a given time frame. There has to be, therefore, a prior assessment of the degree of probability of occurrences.

(ii) Undertaking measures—simultaneously with hazard-zoning—for the reduction of the severity and frequency of hazards, and thus minimizing the extent of resultant damage (Table 7.2).

(iii) Educating people on the nature, causes and consequences of the hazard, the risks involved, and the likely impact on natural systems and the socio-economic structure of society. Dissemination of this information is to be followed by timely warning through various media, with counselling on various hazard-coping measures, including evacuation of the threatened area.

(iv) Mobilizing, in a coordinated manner, relief measures—provision of medicine, good sanitary conditions, food, shelter, etc—required in case of disaster.

[1] The Medvedev, Sponheuer and Karnik scale was first used in 1964. It was used widely in Europe in the mid-1970s, till 1981.

(ii) Non-availability of loans from banks for construction and repair of houses, etc in these zones.

(iii) Denial or withdrawal of essential services/facilities such as water and electricity, public sewers, telephones, buses and taxis, etc from hazard-prone areas.

(iv) Denial of insurance against risks from hazards in hazard zones.

However, there is another viewpoint (Bhandari, per. com. 2002) that is equally valid. In a democratic society, the government must ensure basic amenities to its people. They cannot afford to withdraw these facilities and let the people suffer. If an area is hazard-prone or affected by hazards, the government is obliged to develop the roads faster, develop more hospitals and schools, strengthen the buildings and utility structures, or carry out repairs. Insurance against hazards must be encouraged. This will, of course, encourage the people to remain in the hazard-prone areas and carry on activities that would only amplify risk from hazards.

Hazard-prone areas can be acquired by the government to develop recreation parks, playgrounds, fuel farms, horticultural and vegetable gardens, etc.

Monetary incentives could also work:

(i) Encouraging business people to invest in hazard mitigation measures by providing income tax credits.

(ii) Exempting tax for private bonds that are used in retrofitting buildings and facilities.

(iii) Matching grants from the central government for retrofitting critical facilities belonging to the state and local governments.

(iv) Forming a National Disaster Insurance Corporation that provides insurance and reinsurance.

In addition to the incentives and disincentives provided by the central government, it is necessary that the state and local governments are better prepared to carry out hazard mitigation work, with the help of training and drills. Individuals must be encouraged to assume more active roles in disaster preparedness, response and recovery programmes.

It must be admitted that post-disaster funding is an incentive of sorts, especially if the government bears one hundred per cent of the disaster management costs. It is equivalent to a reward for neglecting prior disaster management planning. It is therefore advisable to put money on hazard preparedness plans rather than on relief and rehabilitation activities.

## DANGER FROM SPACE

Some scientists harbour a fear of events similar to those that occurred 65 million years ago, when a giant asteroid is believed to have struck the earth and wiped out the dinosaurs. There are 200 craters around the world that testify to the fact that extra-terrestrial bodies do hit the earth. The 49,000-year-old Barringer crater in

Arizona (U.S.A.), the Lonar crater in Buldana district of Maharashtra (1,830 m across and 150 m deep, formed less than 50,000 years ago) and the Tunguska disaster in AD 1908, when a meteorite destroyed forests over an area of 160 km in Siberia, remind us that the earth does face dangers from space. Admittedly, the chances of a giant asteroid or a comet striking the earth are not great. But if that happens, the damage will be unimaginably high. For, if a kilometre-wide object were to hit the earth, the energy unleashed would be ten million times greater than the power of the atom bomb dropped on Hiroshima!

Astronomers believe that debris from Encke, a comet that orbits the sun between Jupiter and Mercury, has struck the earth with the violence of a nuclear explosion several times in the past. The most recent impact is believed to have occurred in June 1908 in Siberia. The earth runs into the trail of dust and debris around Encke every June and November (*The Hindu*, 18 June 2002). Swarms of comets and asteroids move around the sun and in the proximity of the earth. Most of these objects are frozen balls of ice and dust and pieces of asteroids and comets. These objects are sometimes driven close to the sun due to gravitational pull, or through mutual attraction. In such cases, they are likely to cross the earth's orbit. Since there are about 1,413 such objects near the earth, the chances of some of them colliding with the earth cannot be discounted.

The Indian Space Research Organization can play a very critical role in spotting such objects. Advance knowledge of areas which they are likely to impact would help prepare in coping with the hazard. If an earth-bound object is spotted, the use of rockets and nuclear missiles would be necessary to deflect it, or destroy it in space, before it reaches the earth.

# 8

# Preparing for Earthquakes and Planning Hazard Mitigation

## The Threat of Earthquakes

More than 100,000 lives have been lost in earthquakes in the recorded history of the Indian people. The annual toll of human lives the world over is about 16,000, of which more than four-fifth have occurred in developing countries. It is sobering to realize that about 85 per cent of the world's population living under the threat of earthquakes belongs to developing countries, including India (Cardona et al, 1998). The loss in economic terms is more than 40% of the Gross National Product in developing countries, but less than 1% in developed countries. Awareness of the risks of seismic activity is also very low in developing countries.

There is an inordinately rapid growth of population in megacities like Dhaka, Kolkata, Lucknow, Kanpur, Delhi, Lahore and Layalpur in the Indo-Gangetic Plain. This means that future earthquakes can cause 5,000 deaths a year on average (Keilis-Borok et al, 1984). Likewise, in the Karachi–Kachchh–Saurashtra sector in western India which has a population growth rate of 3.1 per cent, the likelihood of death from earthquake hazards is more than 8,000 per year. The Bhuj earthquake of January 2001 showed how correct this estimate was. More than 20,000 people died in building collapse resulting from that earthquake.

**Table 8.1.** Loss of human life due to earthquakes in the Indian subcontinent (Arya, 2000).

| Year | Locality | Death Toll |
|---|---|---|
| 1839 (16 June) | Kachchh (Gujarat) | 15,000 |
| 1855 | Badgam (Kashmir) | 3,000 |
| 1897 (12 June) | Shillong (Meghalaya) | 1,600 |
| 1905 (4 April) | Kangra (Himachal) | 19,500 |
| 1934 (15 January) | Northern Bihar | 10,500 |
| 1935 | Quetta (Pakistan) | 30,000 |
| 1950 (15 August) | Northeastern Arunachal | >1,500 |
| 1988 (21 August) | Udaypur (Nepal) | 1,000 |
| 1991 (20 October) | Uttarkashi (Uttaranchal) | >1,500 |
| 1993 (30 September) | Latur (Maharashtra) | >9,000 |
| 2001 (26 January) | Bhuj (Gujarat) | >20,000 |

## IMPACT ON THE ENVIRONMENT

Earthquakes are responsible for the formation of fault scarps and depressions resulting from displacement of the ground along rupture planes. The 1819 earthquake in northern Kachchh gave rise to a more than 90-km long and 2.5–3.7-m high east–west trending ridge, whose scarp face is called the Allah Band (Plate 8A). Fort Sindri sank five to eight metres in the depression that was formed adjacent to this ridge. Similarly, the 1897 earthquake in Meghalaya caused a vertical displacement of 11–30 m along a 20-km north–south striking fault parallel to the Chedrang river. As a result, a number of streams dropped to become waterfalls across this reactivated fault and many were ponded, giving rise to as many as 30 lakes. The 1905 Kangra earthquake uplifted the Dehradun ground by 13.4 cm, resulting in the deflection of many Siwalik streams.

At the point where a **seismogenic fault** crosses the Brahmaputra river north of Guwahati, the normally 8–10 km wide channel narrows down considerably in an abrupt manner, allowing the river a restricted passage across the uplifted block of the Meghalaya massif (Plates 4B and 14). This development is possibly the outcome of the 1897 earthquake. This has caused ponding of the Brahmaputra and deposition of an enormous volume of sediments in the channel (Valdiya, 1999).

The 1950 earthquake in northeastern Arunachal caused a 2.5–3 m uplift of the Brahmaputra bed near Dibrugarh. The river was blocked for a while, and filled rapidly with sediments that came from the landslide-ravaged mountains of Arunachal Pradesh. The result was unseasonal floods in Assam, and ground shaking that caused unstable hill slopes to slide down in multitudes of landslides and debris avalanches. The earthquake generated more than 4,600 million m$^3$ of debris over a 300-km long by 100-km wide area in landslide-ravaged eastern Arunachal Pradesh (Ramachandran Rao, 1953). The Lohit, Dibang and Dihing rivers were dammed at several places, and the debris choked the Brahmaputra channel upstream of Dibrugarh. The bursting of these dams later brought about unseasonal floods in Assam.

Alluvial deposits in plains, and fluvial terraces and lacustrine successions in plains and mountain, are mobilized underground when shaken by large-magnitude earthquakes. Liquified material is squeezed up forming **seismites**, **sand dykes** and **sand sills** underground, and **sand blows** or **mud volcanoes** on the ground surface (Plate 8B). Thousands of mud volcanoes—pouring out clay, silt and fine sand, along with copious water—were active in the Rann of Kachchh and Banni plain after the 2001 Bhuj earthquake (Plate 8B). Similar phenomena were witnessed after the 1934 earthquake in northern Bihar. Thousands of mud volcanoes spouted out sand, mud and water and subsequently submerged the vast agricultural land in knee-deep water, squeezed out of the mobilized aquifer below the surface. It was a deluge, without rain, during a dry winter!

Circulation of groundwater is not the only aspect severely affected by the shaking and squeezing of aquifers. Water levels in wells also change drastically—going down in some places and rising up in others. There is also considerable change in the discharge of springs.

Earthquakes have considerable impact on the environment of biotic communities, including humans. Great earthquakes in the distant past are known to have changed the lifestyles of people, forcing people to migrate, and often changing the course of civilization. Evidence suggests that this happened to the Harappans of the Saraswati floodplains some thousands of years ago (Valdiya, 2002).

## Causes of Earthquakes

When a rock body under compression snaps or breaks (Figure 8.1) and the strained rocks rebound along the plane of rupture, there is an earthquake. This results in one block sliding past or under the other on the plane of rupture—the fault. Due to compositional variations and structural characteristics, rock bodies are commonly heterogeneous. Due to this heterogenity, rupturing is always irregular. The snapping starts at a point, called **focus** or **hypocentre**, and propagates at certain velocities to cover the entire area of the main fault. As the rupture spreads, the attendant tremors and rumbles travel at the rate of 2–3 km/second (always less than 3 km/second). The movement along faults is not continual, rather there is a storing up of the slip, which is later released in frequent earthquakes. The faults that produce most earthquakes are the ones that accumulate slip at faster rates. In the case of long faults, small segments are reactivated at a time, giving rise to smaller earthquakes. Where the fault planes bend, the block motion becomes quite complex.

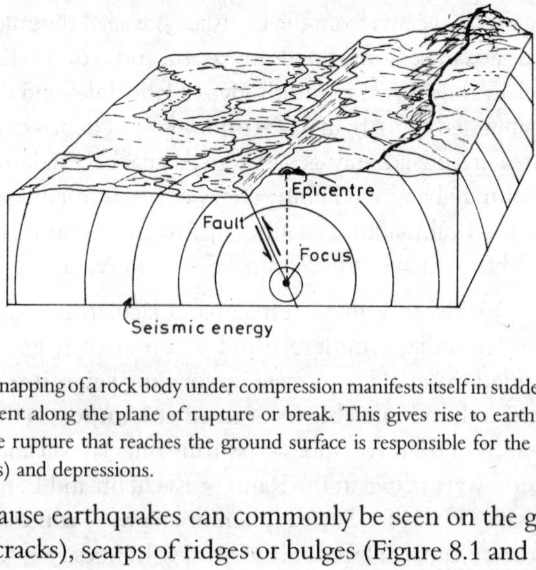


**Figure 8.1.** Snapping of a rock body under compression manifests itself in sudden movement or displacement along the plane of rupture or break. This gives rise to earth tremors and rumbles. The rupture that reaches the ground surface is responsible for the formation of scarps (ridges) and depressions.

The faults that cause earthquakes can commonly be seen on the ground surface as rupture planes (cracks), scarps of ridges or bulges (Figure 8.1 and Plate 8A), and as depressions. The ruptures are never open fissures, for the ground moves across the fault, not away from it, during earthquakes. There are other faults which remain concealed beneath the cover of rocks and/or sediments. These are faults which fail to reach the surface, called **blind faults**. The 1905 earthquake that struck Dehradun is believed to have originated from a blind fault (Yeats and Thakur, 1998). Only detailed geophysical investigations combined with geomorphological-structural studies can reveal the existence of blind faults.

## Propagation of Seismic Waves

The rupture spreads in a series of uneven movements or displacements of rock bodies, generating bursts of high-frequency waves. These are seismic waves that travel in all directions. Seismic waves are of two types—**body waves** and **surface waves**. Among body waves the **primary wave** travels faster, and entails the vibration of particles back and forth in the direction of wave travel, which cause compression and dilation of the rock body. This is manifest in the cracking and deformation of the ground surface. The **secondary wave** or **shear wave** involves left-to-right and up-and-down vibrations of particles in planes perpendicular to the direction of propagation. Secondary waves are slower in speed but cause greater damage, and are responsible for dominant vibrations or ground shaking. The second type of seismic waves, **surface waves**, cause a rhythmic rise and fall of the ground as they pass. The very rapid rise and fall over a period of time—from a few seconds to a couple of minutes—cause instability or damage to structures located on the ground.

Ground shaking depends on the magnitude of the earthquake, the distance from the causative fault and the soil condition. The softer soils and unconsolidated sediments of floodplains, lake succession and fluvial terraces experience greater shaking and resultant liquifaction. Pore-water pressure increases within loosely packed water-saturated sediments or soil when shaken by an earthquake, or subjected to repeated or heavy loading. When the build-up of pore-water pressure exceeds the overburden pressure, the sediment/soil loses its shear strength and behaves like a liquid (Bhandari, 2002). The loose sediment begins to flow and fill fissures and cracks on the overlying material, giving rise to sills, dykes and what is known as **seismite**. It could also come out onto the surface, forming sand boils, sand blows and mud volcanoes (Plate 8B). This was noticed recently (in January 2001) in the Rann of Kachchh when an earthquake of $M_w$ 7.7 struck the region. Sometimes, spouting water may rise as high as two metres, as witnessed during the 1934 Bihar earthquake.

Seismic waves die off as they travel through the earth—the high-frequency body waves dying more rapidly than the low-frequency surface waves. Near the origin of the earthquake, the high frequencies of vibration are experienced appreciably—there is therefore the feeling of a jolt in and around the epicentre. The **epicentre** is the ground above the hypocentre or origin of the quake. Farther away, the high-frequency waves die out and only the lower-frequency vibrations (surface waves) are felt, which give the feeling of a rolling motion.

## Intensity and Magnitude

The intensity of an earthquake is assessed by the degree of damage caused by the earthquake. It depends on the length of the rupture plane (Figure 8.3A). The larger the plane of break, the stronger is the intensity. The largest known rupture plane in the Indian subcontinent was measured at 400 km, in Meghalaya, from the 1897 earthquake of magnitude M 8.7 (Chen and Molnar, 1977). The degree of damage or severity of the earthquake is described in terms of: (i) the observed impact or effects on the environment, (ii) the extent of damage to structures, and (iii) human sensitiveness. This intensity is assessed on the **Modified Mercalli (MM) Scale**. It has twelve divisions (of intensity), each denoted by a Roman numeral (Table 8.2). The intensity varies from place to place—areas of equal intensity are demarcated by curved lines called **isoseismals**.

Ground shaking due to the passage of body waves is recorded on graph paper put on the rotating drum of the seismograph (Figure 8.2A and 8.2B). The amplitude of the waves thus recorded is proportionate to the magnitude of the event that caused the shaking. Charles Richter, in 1935, proposed a scale that would rate earthquakes in terms of the energy released, based on the amplitude of ground motion recorded by the Wood–Anderson seismograph. The amplitude of the surface waves is expressed in the linear scale and the magnitude M of the event in the logarithmic scale (with a range up to 10). This is called the **Richter Scale**. Each ascending whole number of this scale indicates ten times greater amplitude or 31.5 times more energy release than the preceeding whole number. Thus, an earthquake of M 5 has ten times greater amplitude and 31.5 times greater energy release than an earthquake of M 4 (Figure 8.2C).

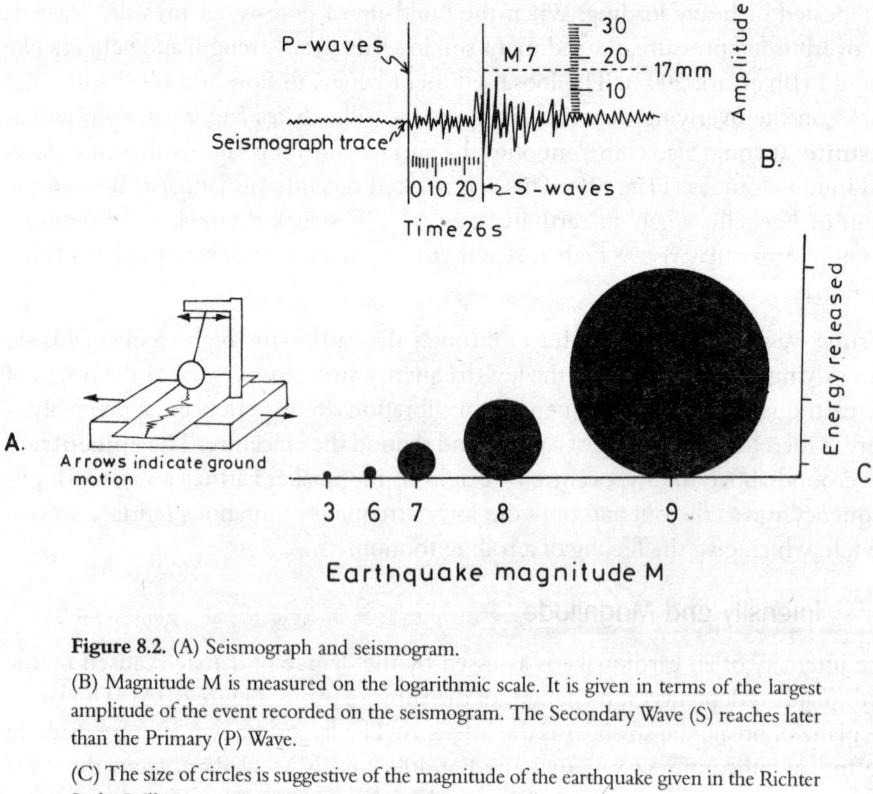


**Figure 8.2.** (A) Seismograph and seismogram.

(B) Magnitude M is measured on the logarithmic scale. It is given in terms of the largest amplitude of the event recorded on the seismogram. The Secondary Wave (S) reaches later than the Primary (P) Wave.

(C) The size of circles is suggestive of the magnitude of the earthquake given in the Richter Scale. It illustrates energy release with magnitude changes.

**Table 8.2.** Measure of intensity of earthquake according to the Modified Mercalli Scale (After Keller, 1976).

| MM Scale | Observation |
|---|---|
| I. | Not felt except by a very few under especially favourable circumstances. |
| II. | Felt only by a few persons at rest, especially on upper floors of buildings. Delicately suspended objects may swing. |
| III. | Felt quite noticeably indoors, especially on upper floors of buildings, but many may not recognize it as an earthquake. Standing cars may rock slightly. Similar to the vibration of a passing truck. |
| IV. | During the day, felt indoors by many and outdoors by few. At night some people may be awakened. Dishes, windows, doors disturbed. Walls make cracking sounds. Sensation is similar to that of a heavy truck striking a building. Standing cars rocked noticeably. |
| V. | Felt by nearly everyone, many are awakened. Some dishes, windows, etc could be broken. A few instances of cracked plaster. Unstable objects overturned. Disturbance of trees, poles and other tall objects sometimes noticed. Pendulum clocks may stop. |
| VI. | Felt by all, many are frightened and run outdoors. Heavy furniture may move. A few instances of fallen plaster or damaged chimneys. Damage slight. |
| VII. | Everyone runs outdoors. Damage negligible in buildings of good design and construction, slight to moderate in well-built ordinary structures, but considerable in poorly built or badly designed structures. Some chimneys broken. Noticed by persons driving cars. |
| VIII. | Damage slight in specially designed structures, considerable in ordinary substantial buildings with partial collapse, and great in poorly built structures. Panel walls thrown out of frame structures. Chimneys, factory stacks, columns, monuments, walls may fall. Heavy furniture overturned. Sand and mud ejected in small amounts. Causes changes in well water. People driving cars are disturbed. |
| IX. | Damage considerable in specially designed structures. Well-designed frame structures thrown out of plumb. Great damage in substantial buildings, with partial collapse. Buildings shift from their foundations. Ground cracked conspicuously. Underground pipes broken. |
| X. | Some well-built wooden structures destroyed, and most masonry and frame structures with foundations destroyed. Ground badly cracked. Rails bent. Landslides considerable from river banks and steep slopes. Shifted sand and mud. Water splashed over banks. |
| XI. | Few, if any, masonry structures remain standing. Bridges destroyed. Broad fissures in ground. Underground pipelines go completely out of service. Earth-slumps and landslips in soft ground. Rails bent greatly. |
| XII. | Damage total. Waves seen on ground surfaces. Lines of sight and level distorted. Objects thrown upward into the air. |

**Table 8.3.** Approximate relationship between Richter Scale Magnitude (M) and Modified Mercalli (MM) Scale intensity (After Keller, 1976)

| Richter Scale (M) | Mercalli Scale (MM) | Distance upto which felt | Ground Acceleration (g) |
|---|---|---|---|
| 3.0–3.9 | II–III | 29 | < 0.15 |
| 4.0–4.9 | IV–V | 50 | 0.15–0.04 |
| 5.0–5.9 | VI–VII | 110 | 0.06–0.15 |
| 6.0–6.9 | VII–VIII | 200 | 0.15–0.30 |
| 7.0–7.9 | IX–X | 400 | 0.50–0.60 |
| 8.0–8.9 | XI–XII | 720 | > 0.60 |

In recent years, a new scale called **Moment Magnitude ($M_w$)** is being used. The moment is a physical quantity related to the total energy released in an earthquake. This is estimated by examining the geometry of the causative fault in the field or by analysing a seismogram in the laboratory. This scale has made possible the measurement of all kinds of earthquakes—big and small, near and distant, recent and past—with the same scale, making it very reliable.

## Earthquake Shocks

Large earthquakes commonly occur in sequence and the epicentres occur in clusters. The strongest quake is called the **main shock**. The ones before the powerful main event are known as **foreshocks**, and those that follow it are called **aftershocks**.

Foreshocks and aftershocks usually occur in the geographic proximity of the main shock. Larger earthquakes are generally followed by aftershocks within the first hour (Jones, 1995). On average, the magnitude of the largest aftershocks are 1.2 units less than that of the main shock, the difference of intensity varying between 1.0 and 3.0. Understandably, as the earthquake-causing stresses are released, the rate of occurrence of aftershocks decreases rather quickly (Figure 8.3B). It follows a power law (proposed by F. Omori in 1894), which states that the number of aftershocks on the second day after the main shock will be half the number of those on the first day; on the third day they will be one-third the number, and so on. The size of the main shock and the geodynamic activeness of the region determine the period over which the aftershocks will continue. It can be for a few days, a few years, or even a few decades. Rarely are the main shocks followed by more powerful earthquakes. 'The most likely time for a main shock is within the first hour of the foreshocks. And after three days the risk is almost gone.' (Jones, 1995).

## Implications of Ground Acceleration

For any object in a stable condition at a given place, the normal value of acceleration due to gravity is 1 'g'. 1 g is measured as 980 cm/second/second at sea level. Vertical up-and-down motion or vibration of particles due to seismic waves acts counter to the normal 'g' at that place. This diminishes the steadiness or stability of the object. Large earthquakes, entailing stronger vibrations, produce greater unsteadiness or instability of the structure. **Peak Ground Acceleration (PGA)** causes maximum instability, which occurs at the epicentral area. The closer an area is to the causative fault, the greater would be the instability and attendant damage. The PGA is estimated on the basis of the geological structure of the area, the geophysical conditions prevailing, the distance from the point of the origin of earthquake and the thickness of the soil or sediment cover.

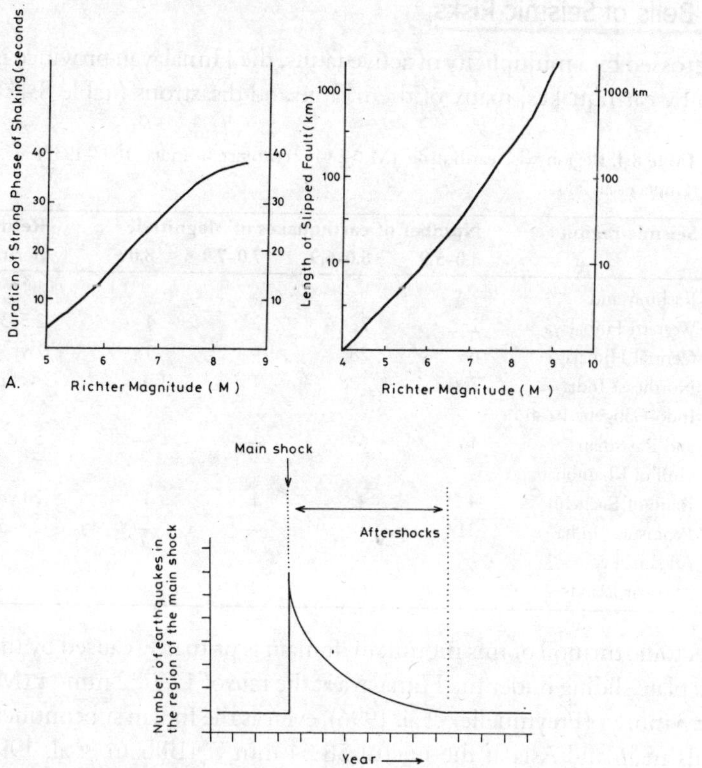


**Figure 8.3.** (A) The duration of strong shaking depends on the magnitude of the earthquake, which in turn depends on the length of the fault plane that slipped.

(B) There is a rapid decrease in the number of aftershocks following the main shock. The time period of the aftershock could be a few days or a few years (After Jones, 1995).

To illustrate this, the 84th percentile PGA at a distance of 15 km from the hypocentre of an earthquake of M 8 would be 0.5 g, if the earthquake were to result from strike-slip movement, and 0.7 g if the displacement took place on a reverse fault. This implies that at a distance of 15 km away from the causative fault, the stability of the structure related to the acceleration due to gravity would be 1 g – 0.7 g = 0.3 g, merely one-third of the original parameter of the stability. Understandably, the structure is bound to topple over when hit by earthquakes of M 8 and above.

A high frequency seismometer (accelerometer) provides a record of ground acceleration over a wide range of frequencies. This is indicated as a fraction of gravitational 'g'. It is generally the practice among engineers to analyse the frequency characteristics (response spectrum) of ground motion. This is useful to design structures which have their own natural frequencies. Spectral analysis is done for a variety of reasons—from determining the source parameters to engineering applications. In simple terms it tells us how different frequency components are associated with different levels of displacement. It is mathematically derived from acceleration data.

## Belts of Seismic Risks

Criss-crossed by a multiplicity of active faults, the Himalayan province is frequently shaken by earthquakes, many of them large and disastrous (Table 8.4; Plate 9).

**Table 8.4.** Region-wise earthquake (M >5.0) occurrence in India (1897–1937) (Gupta et al, 2001).

| Seismic region | Number of earthquakes of Magnitude | | | | Return |
|---|---|---|---|---|---|
| | 5.0–5.9 | 6.0–6.9 | 7.0–7.9 | 8.0+ | period |
| Kashmir and Western Himalaya | 25 | 7 | 2 | 1 | 2.5–3 yrs |
| Central Himalaya | 68 | 28 | 4 | 1 | 1 yr |
| Northeast India | 200 | 128 | 15 | 4 | < 4 months |
| Indo-Gangetic basin and Rajasthan | 14 | 6 | — | — | 5 yrs |
| Gulf of Khambhat and Rann of Kachchh | 4 | 4 | 1 | 1 | 20 yrs |
| Peninsular India | 31 | 10 | — | — | 2.5–3 yrs |
| Andaman & Nicobar islands | 80 | 68 | 1 | 1 | < 8 months |

The tectonic turmoil of this mountain domain is primarily caused by the Peninsular Indian plate sliding under the Himalaya at the rate of 17.7±2 mm/yr (Molnar, 1990) to 20±3 mm/yr (Freymueller et al, 1996), even as the Indian subcontinent converges towards mainland Asia at the rate of 58±4 mm/yr (Bilham et al, 1998). It is this titanic force of compression, produced by the collision of converging continents, that is responsible for strain accumulation in the Himalaya as well as in the framework of the northward-pushing Peninsular Indian Shield (Valdiya, 2001). The abrupt and swift relaxation of the accumulated strain causes rock bodies to snap, reactivates old faults, and the Indian plate to shake violently, often disastrously.

About 50–60% of the Indian landmass is vulnerable to seismic activity of varying intensity. This is particularly strong along the margins of the crustal plates or blocks, where they sink or slide under the adjoining plate or block and along the zones where moving plates were ruptured in the past. In the first category is the **Main Boundary Thrust** with its associated **Himalayan Frontal Fault** at the foot of the Himalaya (Plate 9). In the second category is the **Narmada–Son Lineament** constituting the great geological divide between northern and southern India.

The zone of the Main Boundary Thrust (MBT) is the locus of great earthquakes. Three have occurred in the last one hundred years: 1905 (M 8.6) of Kangra in northwest Himalaya, 1934 (M 8.4) in northern Bihar and adjoining Nepal and

1950 (M 8.7) in northeastern Arunachal. Flattening with depth, the MBT extends northwards as a subsurface detachment plane, between the Himalayan sedimentary pile and its foundation or basement made up of the northerly prolongation of the Indian Shield. Movements caused by the sliding of the Indian plate along this detachment thrust plane are manifest in frequent earthquakes of moderate to low magnitudes in the approximately 130-km wide belt of the inner Lesser Himalaya—along the middle longitudinal terrain that rises in elevation between 900 and 2500 m. This belt of high seismicity (lying nearly 50 km south of the zone of collision of the Indian subcontinent with the Asian continent) is among the most populated regions of the mountain domain.

Some of the significant earthquakes to have struck this region are: 1803 (M>7.5<7.9) of Garhwal, 1833 (M>7.5<7.9) of west Nepal, 1916 (M 7.5) of farwestern Nepal, 1975 (M 6.8) of Kinnaur in eastern Himachal, 1980 (M 6.3) of Dharchula in northeastern Kumaun, 1991 (M 6.6) of Uttarkashi in Garhwal, and 1999 (M. 6.8) of Chamoli in Garhwal. The northwestern and northeastern ends of the Himalayan arc exhibit extraordinarily high seismicity (Plate 9), because the projecting promontories of the northward-pushing Indian peninsula are consistently moving into these parts of the mountain arc.

The northward-advancing Indian Shield has experienced about a dozen earthquakes of M≥6 since the eighteenth century (Ramalingeswara Rao, 2000). Of the many earthquakes that have struck this stable continental region since the fourteenth century, the significant are listed as: 1341 of Kochi in Kerala, 1507 of Nelamangala in Karnataka, 1618 (M 6.9) of Mumbai, 1679 of the Coromandal coast in Tamilnadu, 1764 (M 6.0) of Mahabaleswar in Maharashtra, 1803 (MM IX) of Mathura in U.P., 1819 (M 7.7) of Kachchh in Gujarat, 1827 (MM IX) of Lahore in Pakistan, 1846 (M 6.5) of Jabalpur in M.P., 1848 (MM IX) of Mount Abu in Rajasthan, 1897 (M 8.7) of Shillong in Meghalaya, 1900 (M 6.0) of Coimbatore in Tamilnadu, 1918 (M 7.6) of Shrimangal in Bangladesh, 1919 (M 6.3) Bhavnagar in Gujarat, 1927 (M 6.5) of the Son valley in M.P., 1930 ($M_s$ 7.1/ $M_w$ 6.4) of Dhubri in Assam, 1938 (M 6.2) of Khandwa in M.P., 1956 ($M_s$ 6.1) of Anjar in Gujarat (Kachchh), 1967 (M 6.7) of Koyna in Maharashtra, 1993 ($M_b$ 6.3) of Latur in Maharashtra, 1997 ($M_b$ 6.0) of Jabalpur in M.P., and 2001 ($M_w$ 7.7) of Bhuj in Gujarat (Kachchh). It is noticed that seismicity is largely confined to the linear belts representing faults, shear zones or rift valleys of great antiquity (Figure 8.4). It is obviously the reactivation of older faults and shear zones which gave rise to these earthquakes.

The earthquakes of the southern Indian Shield indicate an overall coherent strain within it, particularly along the shear zones and fault intersections (Talwani and Rajendran, 1991; Talwani, 2000) as evidenced by the earthquakes at Koyna and Latur. The east–west trending arcuate faults of Kachchh have been relaxing the strain stored by a northerly directed compression. The compressive stress is partially transformed into strike-slip motion along the N–S trending faults that define the hidden graben of the Sabarmati plain of Gujarat. A part of the stress is also transferred to the NNE–SSW striking faults of the Aravalli domain, and their northerly

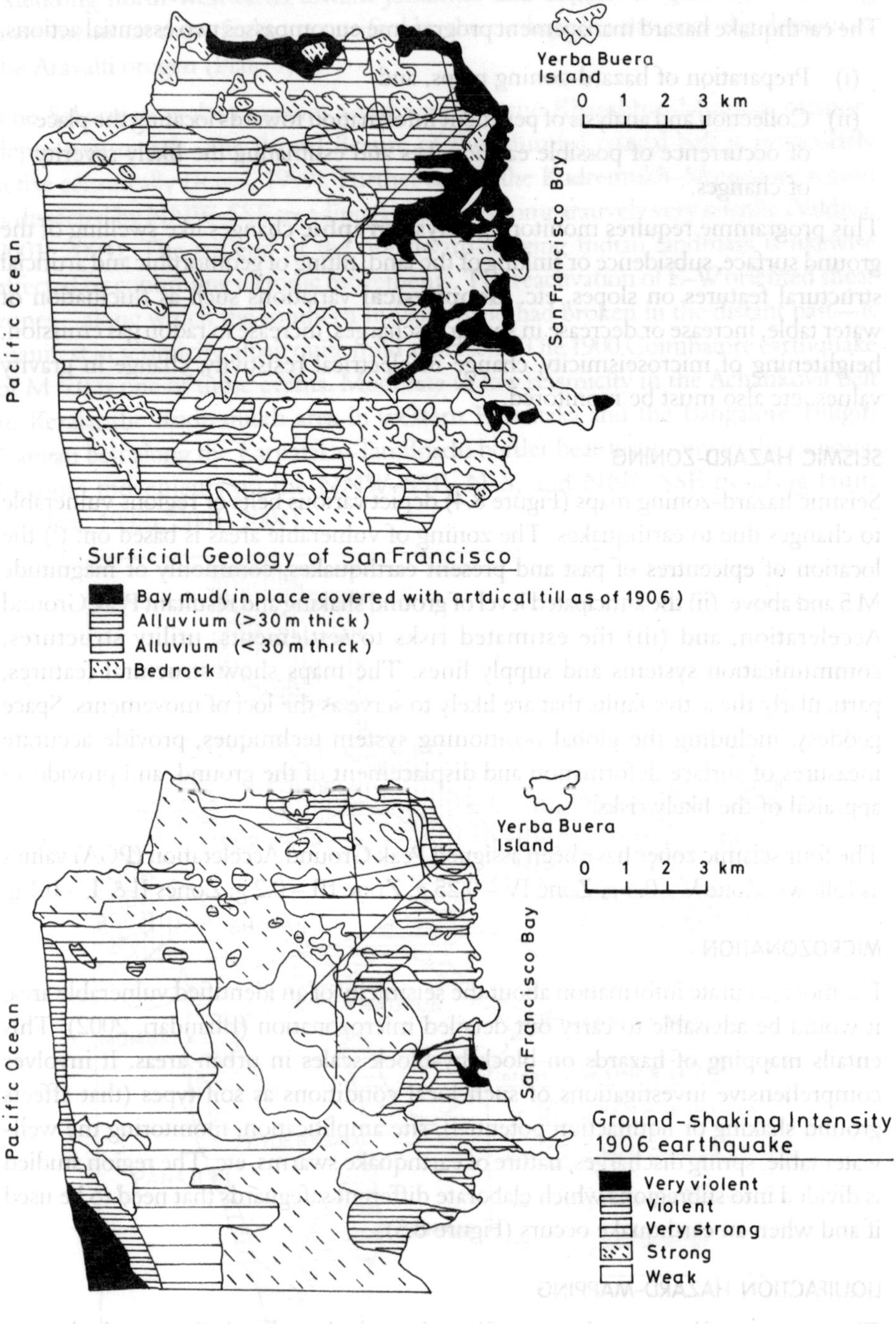


**Figure 8.5.** A model of a microzonation map of the San Francisco region in the USA (Based on Borchendt, 1975).

and bridges/ piers in plains. Lateral spreading of slightly inclined ground causes large deformations of ground surface with its attendant subsidence of settlements. The subsided ground may invite flooding and cause failure of buildings and utility structures, as was seen in Gujarat during the earthquake of 26 January 2001.

It is therefore necessary to undertake identification and delineation of liquifaction-prone areas. The mapping is based on: (i) soil profile, (ii) data analysis, and (iii) value judgement (Bhandari, 2002). It is essential to evaluate the liquifaction potential of the sediments/soils of the regions. This requires forecasting the intensity and duration of the earthquake, and estimating the susceptibility of the ground to liquifaction. The hazard map should show locations of both present and past liquifaction and the potential areas of future liquifaction.

## RECOGNITION OF SEISMIC GAP

It may be emphasized that areas or regions that have been recently ruptured are less likely to be visited by earthquakes in the near future, and therefore are comparatively safer from seismic hazards. However, sectors which have not ruptured are areas where strain is building up progressively without relaxation. These segments are known as **seismic gaps** (Figure 8.6) and are the likely sites of future earthquakes, which will occur when the strain reaches the threshold limit. Looking at seismic events in the period AD 1700–2001, it becomes apparent that:

(i) More than two-thirds of the Himalayan arc (Figure 8.6) remain unruptured (Khattri, 1995, 1999a).

(ii) The periodicity between great earthquakes (M≥8) is 200–270 years (according to some others it is 150–400 years), and

(iii) Approximately 60% of the Himalaya has enough stored elastic energy to drive five to eight great earthquakes (Bilham and Gaur, 2000).

The central sector of the Himalayan domain, embracing western Nepal, Uttaranchal (Kumaun–Garhwal) and eastern Himachal is that part where no great earthquake (M≥8) has occurred since AD 1255. It is this sector that is a 'likely candidate for future earthquakes, the magnitude of which cannot be less than M 8' (Khattri, 1995, 1999a). In this sector, the 1803 Garhwal earthquake and the 1833 earthquake in western Nepal were of the magnitude M 7.7±0.2 (Freymueller et al, 1996). These earthquakes could not have relaxed the strain sufficiently to lower it to safe levels. It is thus a seismically locked sector—a seismic gap (Figures 8.4 and 8.6).

Significantly, major and moderate earthquakes have not occurred periodically or regularly in the highly stressed outer southern belt of the Himalaya (Figure 8.6B, Plate 9). In other words, the accumulated stress along this belt is waiting to be released. The chances of occurrence of one great earthquake in the next hundred-year period is 52% in the central sector, as compared to 27% in the Jammu sector and 21% in the Bhutan–Arunachal gap (Khattri, 1999).

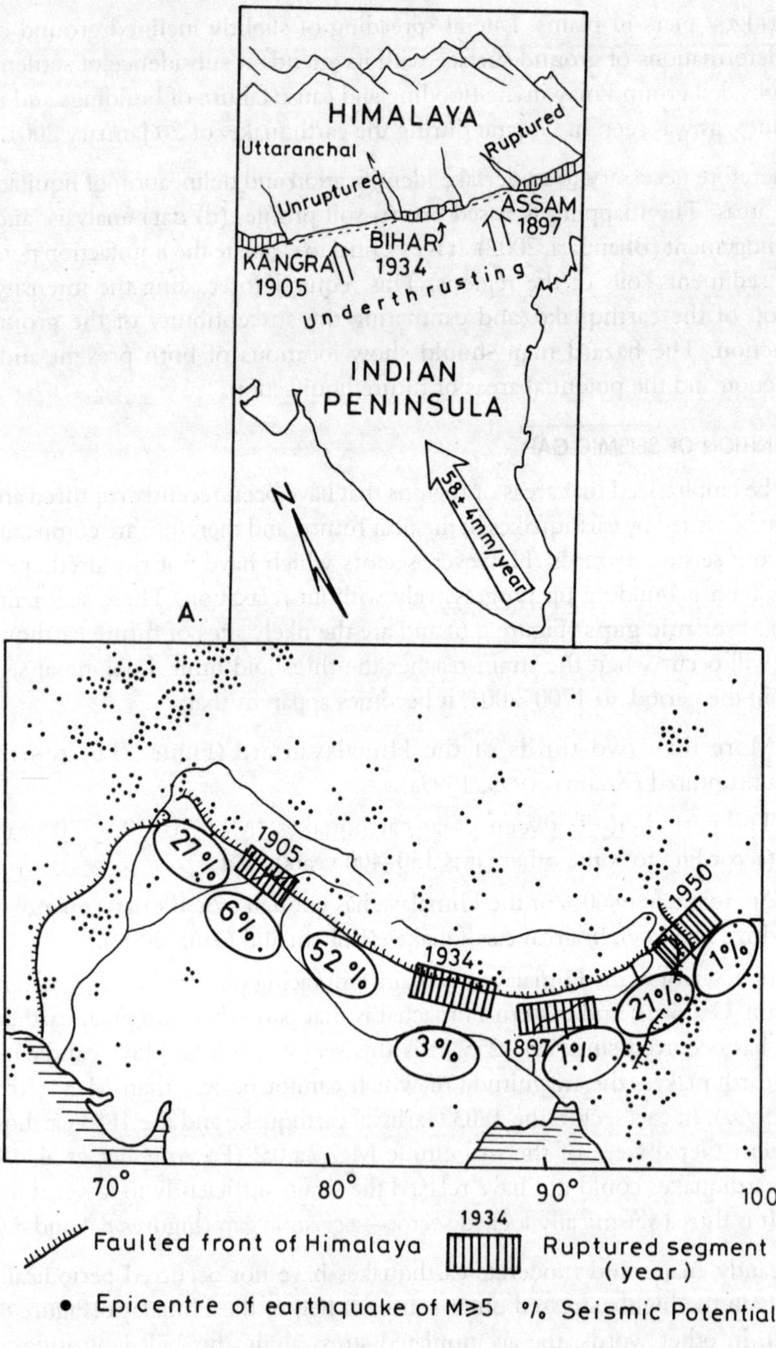


**Figure 8.6.** (A) Conceptual diagram shows the ruptured and as yet unruptured front of the overstressed Himalaya (Gaur, 1992).

(B) Three sectors of the Himalayan arc where no great earthquakes (M ≥ 8) have occurred since AD 1255, are described as seismic gaps (Khattri, 1999a). These are the sectors where great earthquakes are likely to occur.

## Earthquake Hazard Prediction: A Possibility?

It is not possible to predict the time, place and magnitude of earthquakes. This is an unattainable goal despite a few instances where successful predictions have been made. However, it is possible to identify an area likely to be hit by an earthquake in future. This can be done by a critical study of seismic hazard-zoning maps, and rigorous analysis of the trend and the pattern of geophysical and physiographic changes, particularly in seismically locked segments. There are some events that provide clues to probability:

(i) There is a perceptible increase in **microseismicity** before major events (Figure 8.7A). More than 40% of large earthquakes had foreshocks that occurred days and hours before disastrous events (Jones and Molnar, 1976). This was noticed in the case of the 1950 earthquake (M 8.7) in northeastern Arunachal.

(ii) The strain build-up manifests itself in volume expansion of the rock body. This becomes evident from ground swelling and tilting of structures on sloping ground (this can be measured by tiltmetres). The expansion causes development of cracks, even appearing on the ground surface. The expansion may start years before the earthquake.

(iii) There is progressive increase in the quantity of radon gas in pore spaces of soils, in rock fractures, and in well water (Figure 8.7B). Significant increase in radon in groundwater was noticed in the earthquake of 1966 near Tashkent in Uzbekistan.

(iv) The formation of fractures and cracks causes erratic fluctuation of the water table and deformation in the uppermost part of the crust. The water table goes down in some wells or rises in others (Figure 8.8A). In some cases the water level in wells changed by as much as four metres, as was noticed before the 15 August 1950 earthquake in Arunachal. Likewise, spring discharges may change inexplicably, sometimes drastically. This happened before the 1991 Uttarkashi earthquake.

(v) Development of fractures and cracks allows groundwater to seep into the affected rock body and soil. The entry of water causes perceptible decrease in the electrical resistivity (Figure 8.7A) and geomagnetic field of the underground domain.

(vi) Animals and birds are known to behave strangely before impending disasters like volcanic explosions and earthquakes (Figure 8.8B). Cattle, sheep, horses, dogs, cats, rats and snakes seem to have an uncanny perception of coming hazards, and they express their perceptions visibly. Dogs bark untiringly, agitated cattle and horses break away from their tether and run in circles, rabbits jump around with raised ears, rats and snakes come out of their burrows and holes looking quite dazed or stunned, terrified pigeons keep on flying, and so on. The 4 February 1975 Haicheng earthquake of M 7.3 was predicted successfully (Deng et al,

1981), partly on the basis of geophysical changes—particularly foreshocks and their temporal pattern—and partly on the basis of **anomalous animal behaviour** (Figure 8.8B). A day before the 17 January 1995 earthquake, sea lions and hippopotami in the Kobe zoo (Japan) showed strange behaviour—they sank into their pools and would not resurface. Likewise, crocodiles in the Izu-Atagawa garden stopped eating, and silkworms and fish aligned themselves in a common direction before the earthquake. This anomalous behaviour has been attributed to fault movements that produce seismic transient-free charges, by means of a piezoelectric effect of such intensity as to cause lightning discharges (Ikeya et al, 1997).

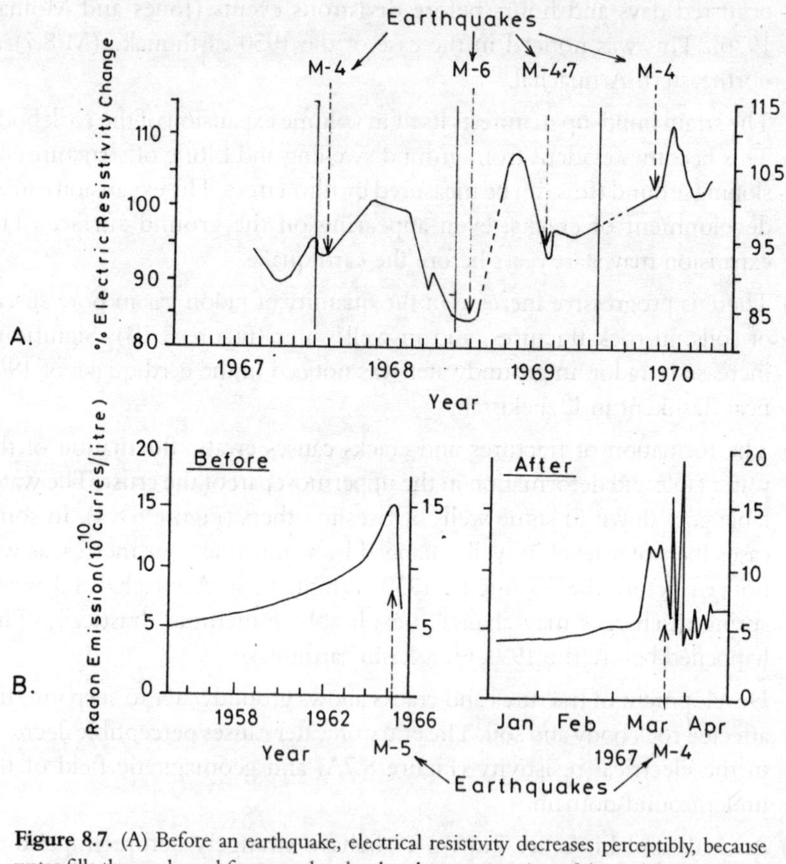


**Figure 8.7.** (A) Before an earthquake, electrical resistivity decreases perceptibly, because water fills the cracks and fractures that develop due to expansion of the rock body.

(B) There is a progressive increase in the amount of radon gas emission before the main shock. (After Keller, 1976).

All these developments can be considered—as the Chinese did to predict the 1975 earthquake at Haicheng, a city of nearly one million people. But the Tangshan earthquake (M 7.8) in 1976 could not be predicted. It killed nearly 240,000 people. Likewise, the American experiment at Parkfield, in the very active seismic belt of California, failed.

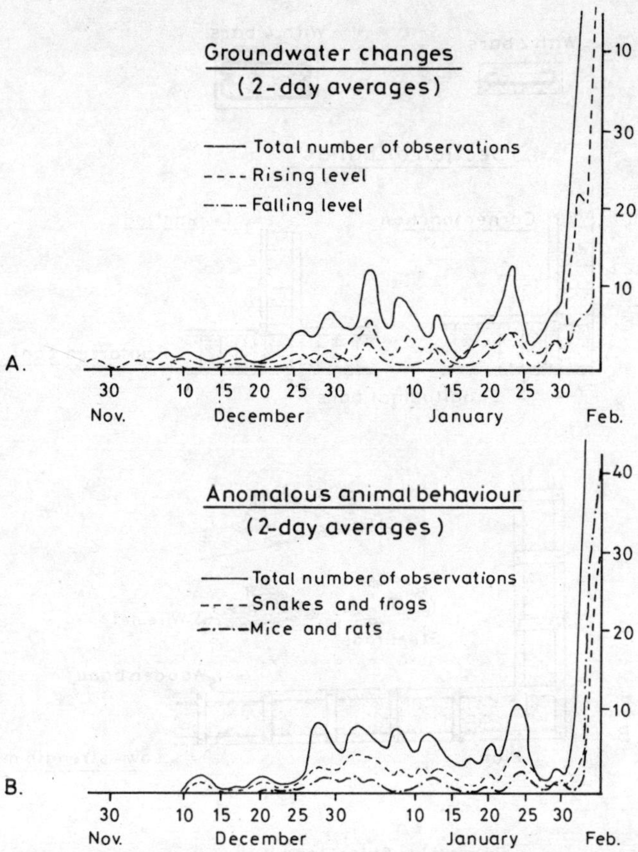


**Figure 8.8.** (A) Before the 1975 earthquake of M 7.3 at Haicheng in China, there was erratic fluctuation of water table over a large area.

(B) Animals and birds behaved strangely before the earthquake of 4 February1975 (After Deng et al, 1981).

## PREPARING TO WITHSTAND THE HAZARD OF EARTHQUAKES

Earthquakes themselves do not cause deaths. It is the events that follow or are triggered—such as landslides, debris avalanches, ground subsidence, floods and building collapse that are responsible for loss of life and property. House collapse is the biggest cause of human casualty. It is therefore necessary that adequate attention is paid to designing and constructing houses that can withstand earthquake shocks.

The vertical component of earthquakes causes vertical acceleration, showing large amplification for short periods. The vertical component enhances tensile stresses in buildings, causing cracking of walls and floors when combined with horizontal acceleration (Arya, 2001). In order to ensure the safety of buildings it is necessary that: (i) the tensile strength and shearing strength of walls be high, (ii) there is an unbreakable bond between walls and corners and T-junctions of walls (Figure 8.9), and (iii) floors and roofs are stiff and connected to the walls, as they serve as a medium to distribute loads (Arya, 2001).

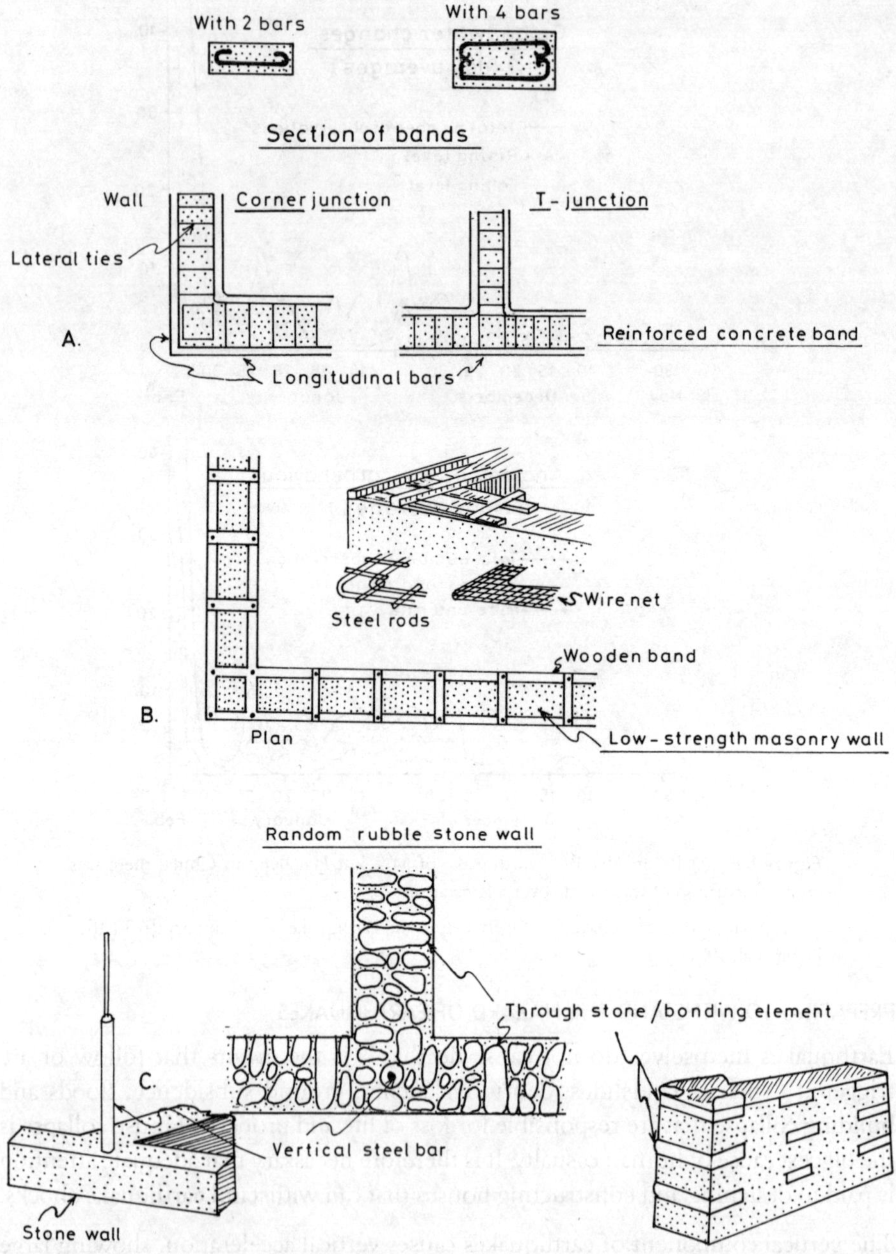


**Figure 8.9.** (A) Reinforced concrete bands at the level of lintel or foundation (or both) comprise longitudinal steel bars held together by lateral ties. The insets above show the sections of the band.

(B) Low-strength masonry walls are strengthened by wooden bands, steel rods or wire nets constructed in the manner shown.

(C) Random stone walls are strengthened by placing wooden/steel bonding elements and long stones across the whole thickness of the walls, and by vertical steel bars. (Based on Paul, 2004).

It is better to desist from making buildings that: (i) are asymmetrical in plan and elevation, (ii) have long and large rooms (rooms larger than 6 $m^2$), (iii) have unduly large windows (more than 30% of the wall length), and (iv) have superfluous projections (overhangs) and cantilevers (Paul, 2004). Buildings can be strengthened to withstand shocks by constructing band beams or bands of reinforced concrete at lintel level (Figure 8.10), and by positioning steel or wooden dowells into the walls at corners or T–junctions (Figures 8.9 and 8.10A, B). Roof trusses are stronger than gable walls (Paul, 2004).

The following measures will help limit damages that are inevitable:

(i) Houses should have solid foundations with a continuous perimeter. This is far better than having structures on pillars and piers. The structures must be fastened to the foundations through bolts or reinforced concrete pillars (Figures 8.9C and 8.11A) so that there is no displacement of the structures when the ground shakes. It is best to have the foundations deeper than 0.3 m and the width more than the height.

(ii) A structure built on grids of reinforced pillars and concrete beams (Figure 8.10)—or some form of horizontal timber coursing (Figures 8.9A, B)—will have sufficient rigidity to withstand ground shaking.

(iii) When the walls are tied up by wall plates (Figures 8.9 and 8.10), and the masonry is reinforced by the use of steel or some wooden components, the structure becomes strong and resilient.

(iv) The building is able to resist damage to a considerable extent if the load of the roof is transferred to the ground, independent of the walls.

(v) If there is a basement or a scrawl space in the structure, it will be advisable to brace the cripple wall around the empty space. This will not allow lateral push when the ground is shaken (Jones, 1995).

(vi) Old houses need to be retrofitted in accordance with their designs (Figure 8.11B).

## PERSONAL SAFETY

In earthquake-prone zones, it is imperative to practise safety measures that will prevent loss of property and even loss of life. Homes, schools, hospitals and work places can be made safer by securing various things with the floor, walls or heavy tables using hooks, fasteners, buckles, safety straps, etc. This is done to prevent the objects from falling or sliding. Window panes and glass partitions need to be covered with strong mylar film so that splinters do not fly. Ceiling fans can be tied to the roof with an extra cable which must be kept slack enough to permit swaying when shaken. Gas appliances should have flexible connections in order to reduce risk of fires if and when the pipes snap.

It is necessary for every member of the family to know the safest place in the building or room. In the event of an earthquake, it is advisable to remain under these safe places, or duck under sturdy tables or desks and remain there until the shaking stops. Outer walls, windows, facades and architectural overhangings are unsafe places and must be avoided. If outdoors, one must avoid power lines, trees, hoardings, big buildings, bridges or overpasses. An important item of the hazard-preparedness programme is making contact with persons outside the disaster-struck area in order to relay messages or seek relief.

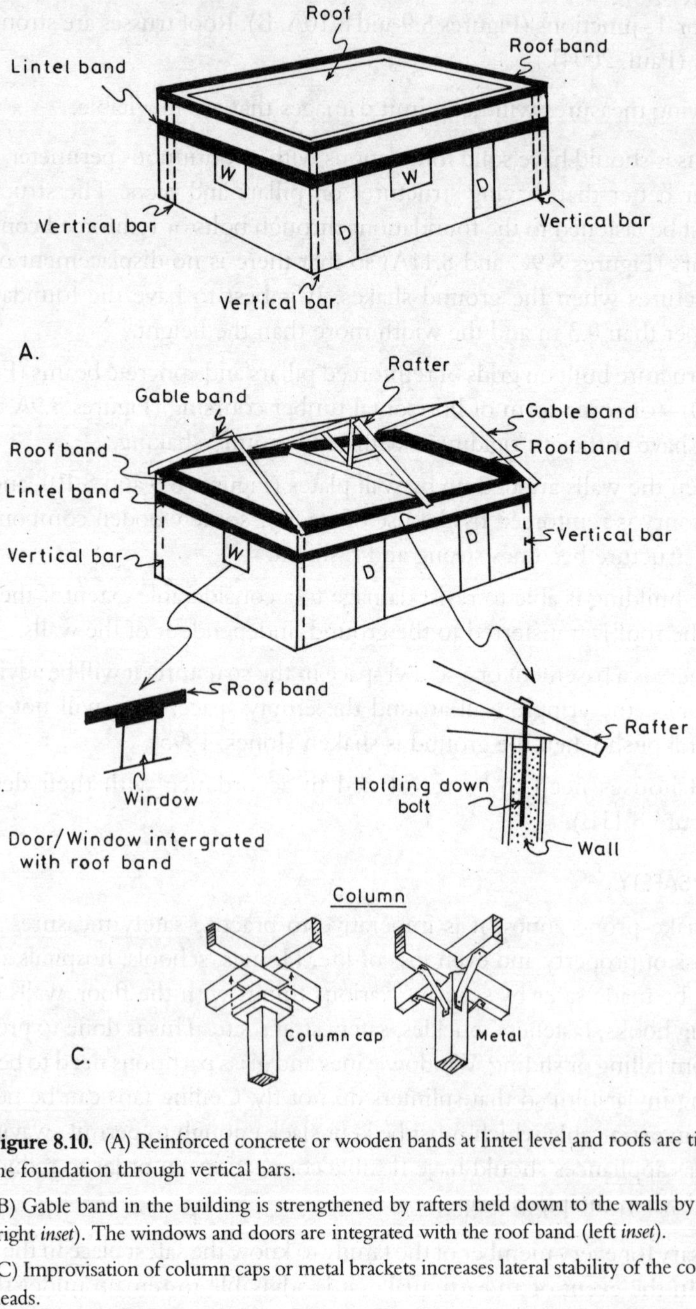


**Figure 8.10.** (A) Reinforced concrete or wooden bands at lintel level and roofs are tied to the foundation through vertical bars.

(B) Gable band in the building is strengthened by rafters held down to the walls by bolts (right *inset*). The windows and doors are integrated with the roof band (left *inset*).

(C) Improvisation of column caps or metal brackets increases lateral stability of the column heads.

(A & B after Paul, 2004; C from Valdiya, 1987)

**Figure 8.11.** (A) Grid of reinforced pillars and beams impart rigidity to the structure. (B) The building frame (superstructure) is bolted to the foundation as a measure of retrofitting (Jones, 1995).

# 9

# Stabilizing Hillslopes and Controlling Landslides

## THE TOLL

In the geodynamically active Himalaya and Sahyadri domains, repeated reactivation of thrusts and faults have weakened the rocks and rendered the landscape vulnerable to stress. Hence, slope failures and resultant landslides with debris flow are quite common during heavy and/or prolonged rains. Where the hillslopes are in 'a critical condition on the margin of their stability' (Haigh, 2000), landslides occur regularly and are hazardous.

In Nepal, landslides leave trails of destruction year after year, the average human toll being 350 per year (*ICIMOD Newsletter*, No. 39, 2001, p. 20). Between 1983–2000, in Nepal, more than 2,300 heads of livestock have been lost, and over 6,700 houses and 4,800 ha of cultivated land damaged by landslides and their attendant debris flow. The condition in the Darjiling–Sikkim Himalaya and the Uttaranchal sector is no better. In Ratnagiri district in Maharashtra, high rainfall coupled with the construction of highways and railway lines, extensive deforestation and hillside housing development is responsible for landslides in slopes steeper than 35 degrees (Thigale and Khandge, 2000). On 9 November 2001, at Neyyatinkara, 45 km north of Trivandrum in Kerala, continuous heavy rain brought down a 700-m hillside, resulting in massive debris flow that swept away houses, killing 36 persons. The rockfalls at Malpa in Kali valley, in August 1998, killed 221 people.

Landslides and their attendant debris flows are the second biggest killers in mountainous domains (after earthquakes) and occur routinely year after year.

## VULNERABLE ZONES

Very large parts of the fault-riven Himalayan province, and some sectors of coastal Kerala and Maharashtra, are prone to landslides during the monsoon when heavy and/or prolonged downpours drench the steep hillslopes. The fault-delimited belts give way during the onslaught of rains. Reactivation of active faults and their attendant seismic shocks and tremors are the real causes of mass-movements during rains. In the Himalayan province, the wide shear zones related to the Main Boundary

Thrust and the Main Central Thrust are characterized by a multiplicity of faults and shattered/crushed rocks. Therefore, landslides are not only frequent but also quite large in dimension (Plate 10). The zone of the Himalayan Frontal Fault at the foothills of the Siwalik, and the Trans-Himadri Fault in the inner Himalaya have also become very vulnerable to landslides, particularly where human intervention has destabilized the mountain sides.

In the Sahyadri–Western Ghat domain of Peninsular India, mass-movements are quite common—particularly in the Thane–Raigarh–Ratnagiri sector in Maharashtra, and southern Kerala. These sectors are cut by NNW–SSE trending faults and fractures, and E–W striking shear zones. With rainfall in excess of 400 mm/yr (between 200–700 mm/yr) and considerable human intervention—due to construction of roads and railway lines, deforestation for water resources development projects, and coffee and tea plantations—the hillslopes, made up of a thick mantle of laterite and/or regolith, succumb quite frequently (Figure 9.1) (Thigale and Khandge, 2000; Thampi, 1997).

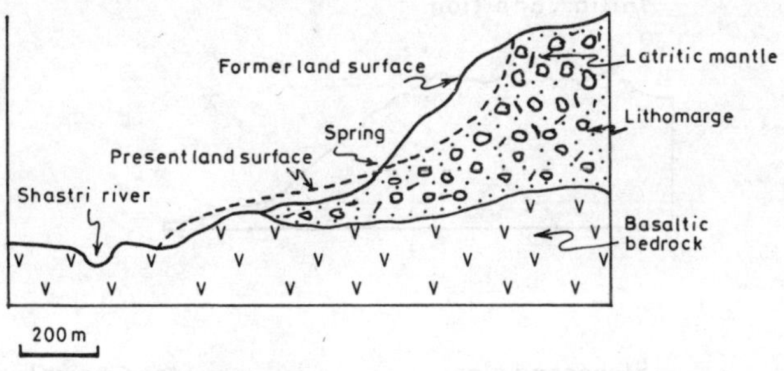


**Figure 9.1.** Landslides occur in the escarpment of the northern Sahyadri (such as at Sangameshwar) where the basaltic rocks wear thick mantles of laterite intimately associated with lithomarge clay (Thigale and Khandge, 2000).

## Triggering of Slope Failure

The forces operating on a hillslope include: (i) the weight of the vulnerable rock and/or soil mass, plus the load of trees, debris dumps, artificial structures, etc, and (ii) the shear strength of rocks/soils counter-acting the slipping (Figure 9.2).

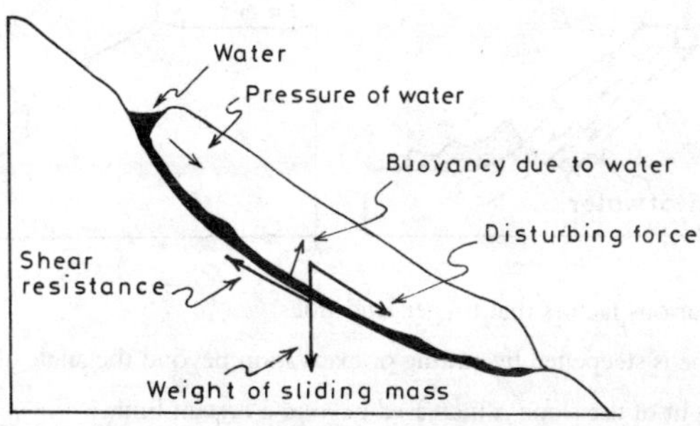


**Figure 9.2.** Forces operating in a slope prone to instability. Water is the main culprit as it diminishes the shear resistance along the plane of slipping.

While the weight acting vertically downwards is the principal cause of the instability, the shear strength—operating along the plane of potential slip—offers resistance to movements, diminishing the chances of downhill slide. The following are the factors that contribute to the downslope movements of vulnerable rock/soil mass:

(i) The entry of water through openings, fissures, fractures, joints, etc reduces the shear strength of the rocks/soils as a result of pore-water pressure. With resistance to sliding thus lessened, the vulnerable mass slips and gives rise to landslides during or after heavy and prolonged rainfall. Towards the end of the rainy season, when the groundwater table is close to the ground surface and the rock/soil mass is very wet, heavy and continuous rains may trigger landslides. Absence or disruption of surface drainage and existence of underground channels greatly promote slope instability (Figure 9.3). Human activities that impede surface flow and disrupt underground circulation of water are also responsible for the building up of pore-water pressure, leading eventually to landslides.

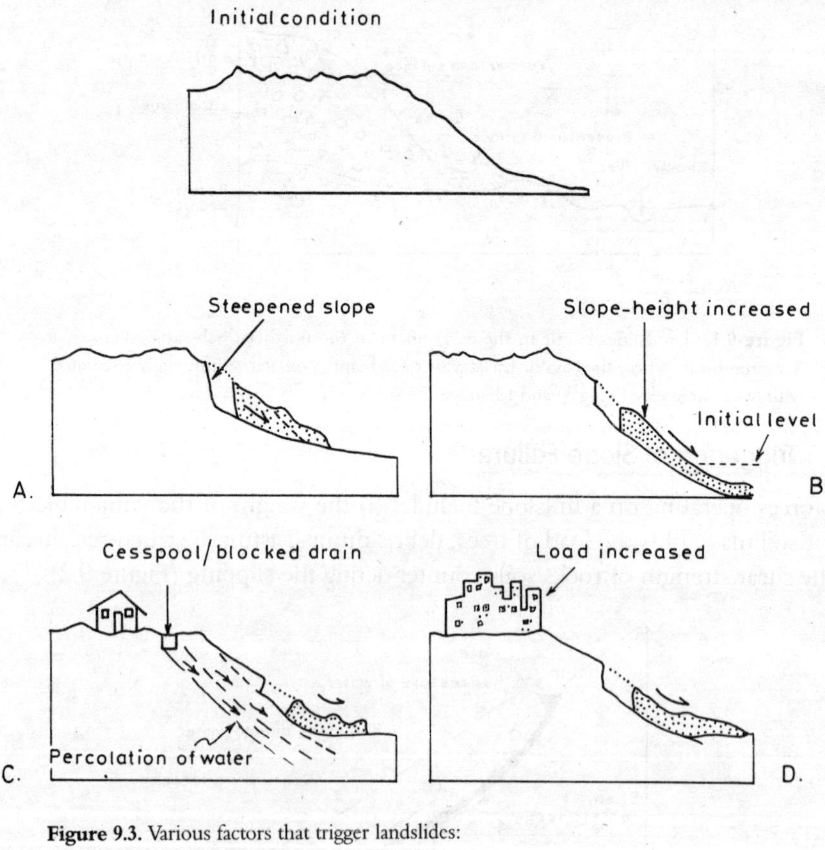


**Figure 9.3.** Various factors that trigger landslides:

(A) If the slope is steepened by cutting or excavation beyond the angle of repose.

(B) If the height of the slope is increased beyond a certain limit.

(C) If water finds its way through fissures, joints or cracks.

(D) If a load is placed atop the vulnerable slope.

(ii) The angle of the slope is an important factor in matters related to its stability. The steeper the slope, the greater is the chance of it becoming unstable (Figure 9.3). Cutting of slopes for roads, railway lines, canals, buildings and mineral extraction steepens the slopes beyond the angle of repose. This promotes landslides.

(iii) Positioning loads like debris mass, buildings and utility structures, big dams, or reservoir water increases the driving force (Figure 9.3), culminating in slope failure. Debris generated by road cutting and mining, if allowed to pile up on the slopes, adds to the vulnerability of the hillslopes. Unplanned buildings have contributed greatly to the insecurity on slopes in the township of Nainital in Kumaun and Shimla in Himachal, as is evident from the continuing creep of the slopes (Figure 9.5).

(iv) Lake or reservoir water exerts lateral (sideways) pressure on hillslopes. Sediments accumulated along the rim of the water bodies serve as buttress walls, contributing to the stability of the hillslope. Withdrawal of lateral support by lowering the water level engenders slope instability, particularly when the water level is lowered suddenly and frequently. The situation becomes serious if there is frequent fluctuation—if the water table rises during the rainy season, and falls low during drier months (Figure 9.4).

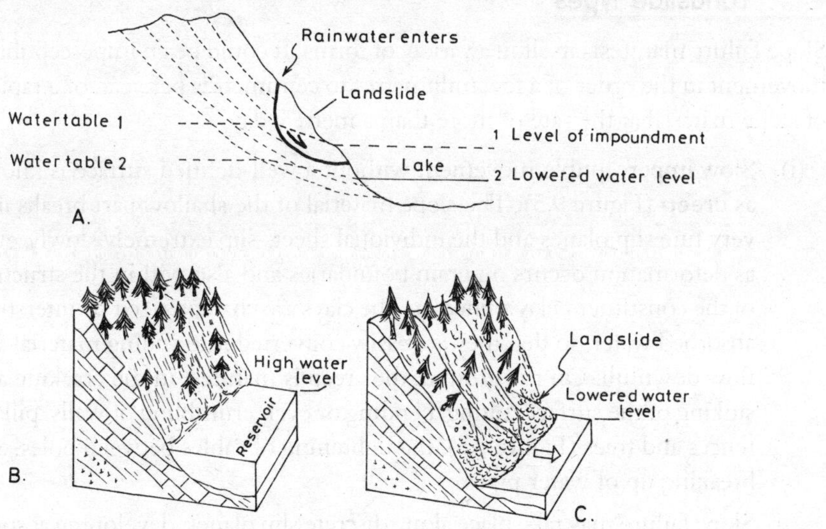


**Figure 9.4.** Withdrawal of lateral support of lake water due to lowering of the water level is instrumental in slope failure.

(v) On a regional scale, large-scale slope failures have been caused by strong earthquake shocks. The 1950 earthquake in northeastern Arunachal (M 8.7) stripped the slopes of the Mishmi range, bringing down 4,600 million m³ of debris down into the valleys of the Lohit, Dibang and Dihing river systems of the Brahmaputra basin. During the 1905 Kangra earthquake (M 8.6), the southern face of the Dhauladhar range crumbled disastrously. The township of Dharmasala, among others, was overwhelmed by rock falls and landslides. Made up of blocks and boulders falling helter-skelter, the multitude of debris fans at the foot of the Dhauladhar bear eloquent testimony to the disaster that followed the earthquake. The Chamoli earthquake of 1999 in Garhwal caused extensive landslides.

The vibrations generated by the passage of heavy vehicles such as trains, tanks, trucks and buses, and blasting for road-cutting, create oscillations of a different frequency in the rocks, in the underground drainage system and in the shear strength of the slope material. These developments affect the forces operating on the slopes. Understandably, such activities tend to accelerate the process of slope failure by widening joints, fractures, fissures, and reducing cohesion of the slope material.

The magnitude–frequency relationship of slope failures, according to Haigh (2000), conforms to a power law that demonstrates that the 'slope system at criticality is characterized by: (i) a sensitive dependence on initial conditions, such that very minor events can set off a chain reaction, (ii) the size and frequency of these events are statistically distributed because they are fractal, and (iii) the distribution holds good in time as well as space.'

## Landslide Types

Slope failure manifests itself in a variety of forms. It could be an imperceptibly slow movement in the order of a few millimetres to centimetres per year, or a rapid slide of slope material at the rate of more than a metre a day.

(i) Slow imperceptible movement without a well-defined surface is known as **creep** (Figure 9.5). The slope material of the shallow part breaks into very fine slip planes and the individual sheets slip extremely slowly, even as deformation occurs on grain boundaries and also within the structure of the constituent clay minerals. The clays are characterized by interstitial absorbed water, so that they are easily converted into sliding material and flow downhill. On the ground, the creep is manifest in the cracking and sinking of the surface, tilting, bending or even crumbling of walls, pillars, fences and trees (Figure 9.5), the tightening of cables between poles, and breaking up of water pipes.

(ii) Slope failure may take place along discrete slip planes, developing at some depth, so that the whole block of rocks and/or soil mass slides rapidly downslope. It can happen along a planar surface (**sheet slide**), or on a concave surface (**slump**). Slumps (Figure 9.6A) are very common where cohesive soils and/or mud, or highly shattered, fissured or jointed rocks make the slope material. These slips are quite damaging and difficult to control. These are far more common where the loose material rests on hard competent rocks, as was noticed in the Mandakini valley in Garhwal following continuous heavy rains in August 1998. A number of villages

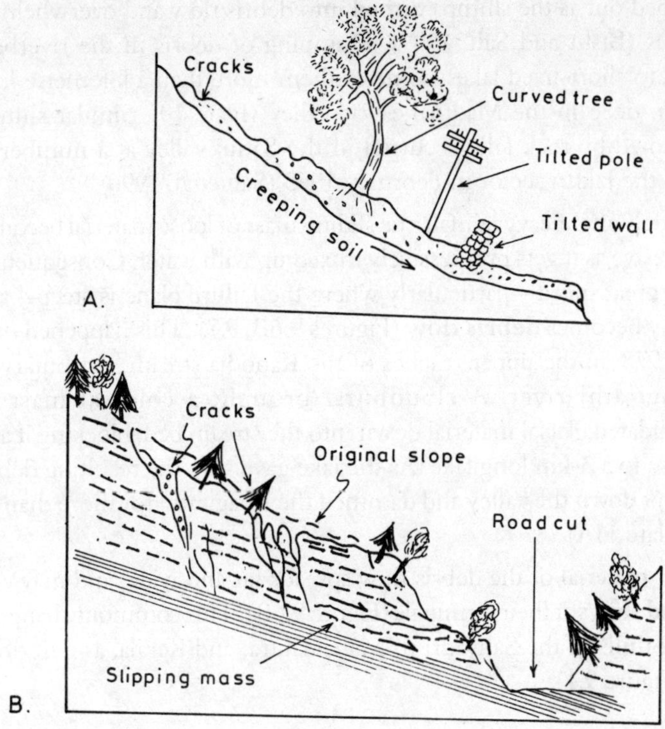


**Figure 9.5.** Slow movement, of the order of a few millimetres to centimetres per day, of superficial material is called creep. It is manifest in cracking of ground and tilting of pillars, trees, etc.

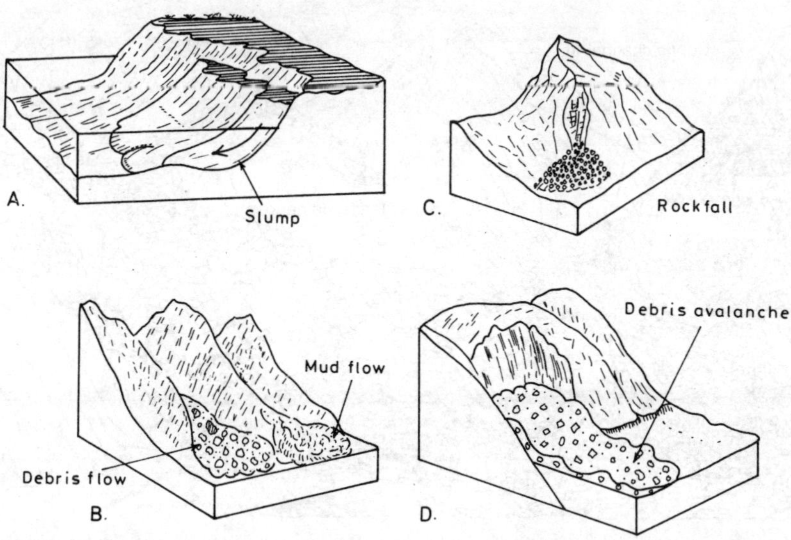


**Figure 9.6.** Types of landslides:
(A) Slump (B) Debris flow (C) Rock fall (D) Debris avalanche.

were wiped out as the slumps turned into debris flow and overwhelmed the slopes (Bisht and Sah, 1999). Dumping of debris in the riverbeds gave rise to short-lived lakes, some of them more than a kilometre long and 30 m deep in the Madmaheswar valley (Plate 10). Similar slumps accompanied by rock falls occurred in the Satluj valley at a number of places in the Jakhri sector in February 1993 (Sah et al, 1996).

(iii) Immediately after heavy rainfall, the sliding mass of loose material becomes non-cohesive as it gets progressively mixed up with water. Consequently, it attains great speed—particularly where the failure plane is steep—and eventually becomes **debris flow** (Figures 9.6B, 9.7). This happened on 6 August 1978, in the upper reaches of the Kanodia stream, a tributary of the Bhagirathi river. A cloudburst brought a colossal mass of unconsolidated glacial material down into the stream bed, blocking it and giving rise to a 3-km long lake. As the lake gave way, the resulting debris flow swept down the valley and dammed the Bhagirathi for more than 14 hours (Plate 11A).

(iv) The finer material of the debris flow spreads far and wide, and may end up in **mud flows** at their terminals (Figure 9.6B). This commonly happens in the foothills of the Sahyadri in Maharashtra, and Kerala, as described earlier (Figure 9.1).

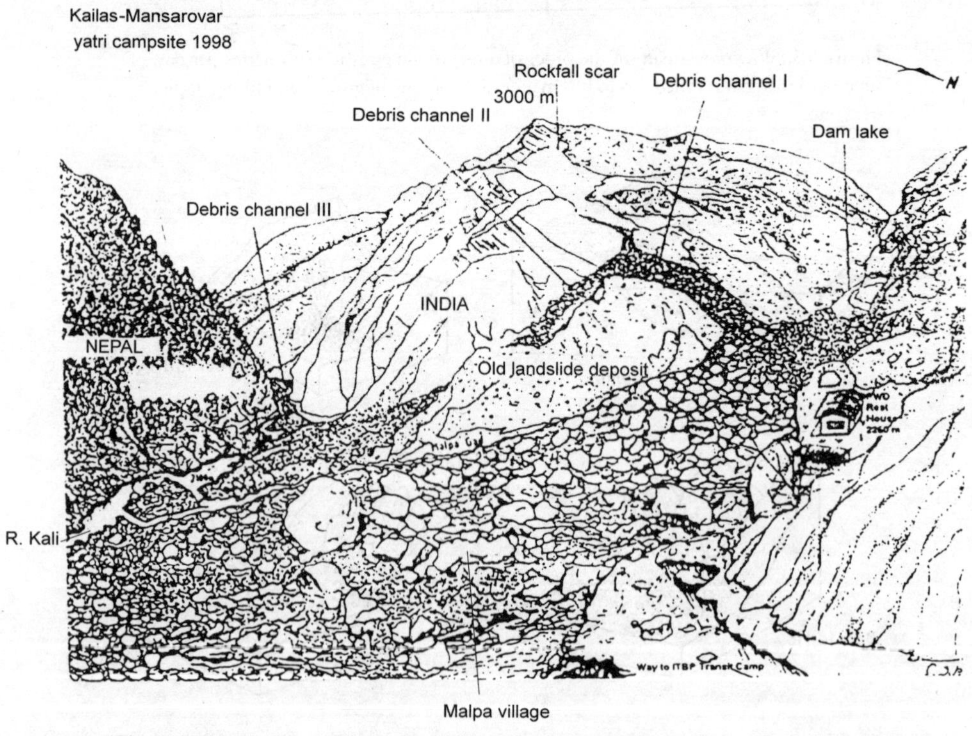


**Figure 9.7.** Rock falls and debris flow from the mountain slope, made up of quartzite and sericite schist, buried 221 pilgrims on the way to Mount Kailas. They were camping at Malpa in the Kali Valley (Paul et al, 2000).

(v) Rains over highly jointed, fractured and sheared rocks often loosen blocks, which may fall from steep slopes. The detached blocks descend, mainly falling, bouncing and rolling through the air. The movement in this case may vary from very rapid to extremely rapid. This is called **rock fall** (Figure 9.6C). In very cold terrains, freezing of water in joints and fractures causes expansion and dislocation of blocks, which leads to rock falls. The Khuni Nala section along the Jammu–Srinagar highway is struck by recurrent rock falls from the Pir Panjal range. On 18 August 1998, the rock falls and debris avalanche from a high mountain slope overlooking Malpa (Figure 9.7) in the Kali valley (northeastern Kumaun) took a toll of 221 pilgrims camping on the river terrace, on their way to Mount Kailas (Paul et al, 2000).

## Disaster Mitigation Measures

### CRUCIAL FIRST STEP

The first step in landslide disaster management is the preparation of hazard-zoning maps showing areas with different degrees of anticipated risks. Satellite imagery obtained from remote-sensing satellites has proved to be very reliable in identifying locations of instability (Plate 12). An example from the Bhagirathi valley in Garhwal eloquently demonstrates the importance of this exercise (R.P. Gupta et al, 1999). Commonly on a scale of 1:50,000 or 1:25,000, the map shows: (i) past and present locations of landslides and debris flows, (ii) lithological variations and the structural setting of the area, (iii) geomorphological features that have a bearing on hillside stability and are indicative of signs of impending failure, (iv) patterns of surface drainage and groundwater conditions, including locations of springs and seepages, (v) land use patterns, including areas of misuse or overuse, etc.

This measure should be complemented with a programme of thorough instrumental monitoring of creep, swelling and subsidence, cracking of ground, and tilting and crumbling of poles, walls, trees, etc.

Hazard-zoning maps and analyses of instrumental monitoring are the basis when assessing risks of landslides in a studied terrain (Figures 9.8, 9.9 and Plate 12). Comprehending the degree or extent of risks should provide guidance to people in the matter of building houses, utility structures, roads, bridges, canals, dams, etc.

### ENSURING SLOPE STABILITY

Water being the agent that initiates and accelerates movement, it is necessary to prevent its entry into the threatened or affected area. This may not stop the event, but will certainly minimize the hazard-potential and severity. The entry and activity of water can be prevented by constructing efficient networks of surface as well as underground drainage. Drainage should be constructed both within and around the affected areas. The drains and ditches are to be properly lined by slates or polypropylene sheets, (Figure 9.10) and they must have sufficient gradient to let

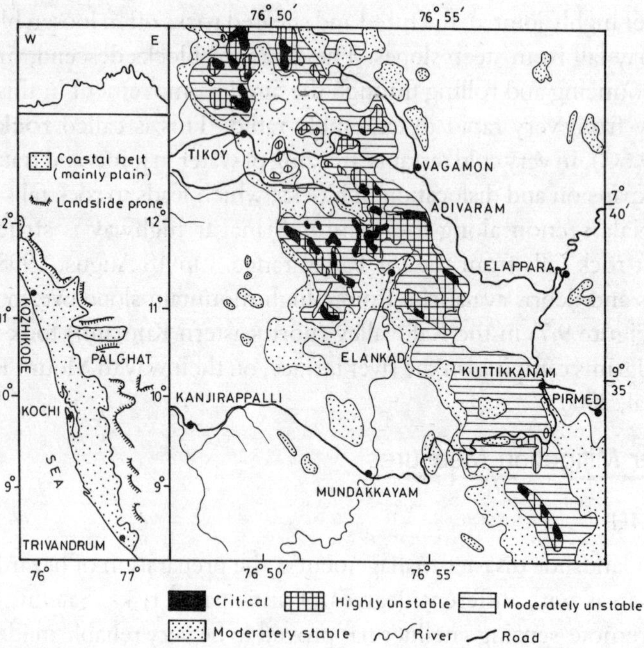


**Figure 9.8.** Landslide hazard-zoning map of a part of Kerala, in the foothills of the southern Sahyadri. It shows zones of various degrees of anticipated risks related to landslides and debris flows (After Thampi et al, 1997).

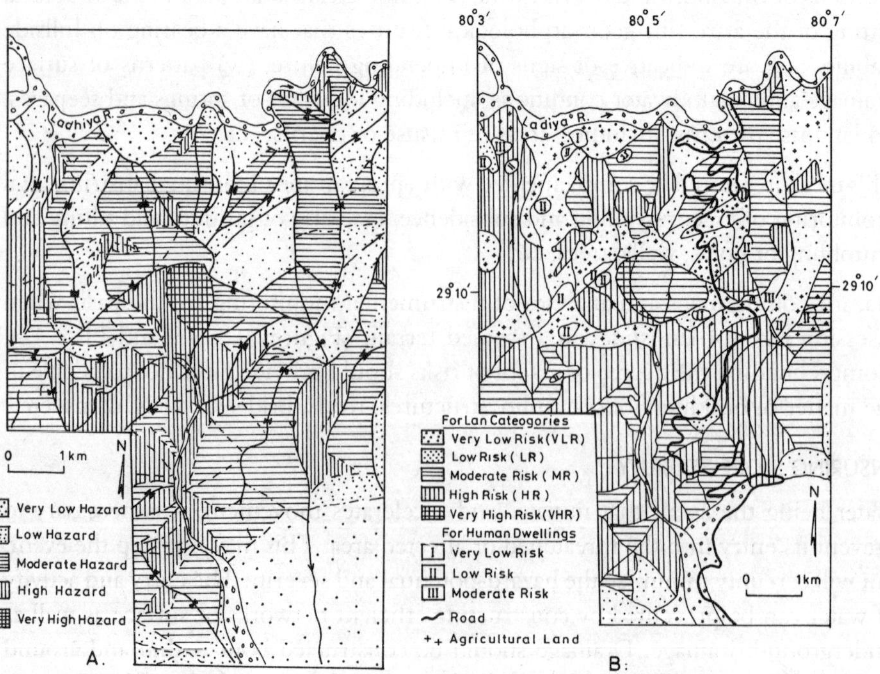


**Figure 9.9.** Landslide hazard-zoning and risk assessment in the Siwalik near Sukhidhang in southeastern Uttaranchal (Anbalagan and Singh, 2001).

the water flow without stopping. They can be filled with gravel or coarse rubble to prevent clogging by garbage and debris slipping from sides. In the Tista valley in Sikkim, horizontal or reclined perforated pipes and bamboo poles were used to drain out water from inside the slid mass. Removal of water, or **dewatering**, relieved the affected slope from the build-up of pore-water pressure (Figure 16.4B).

A more costly way of tackling the problem of water is to cover the threatened area with impermeable layers of clay or asphalt. The cracks and openings on the slopes are plugged—or grouted—with cement, hydrated lime slurry, sodium silicate solution, or bitumen emulsion. A complementary measure is to entrap springs and seepages and divert their discharges through pipes.

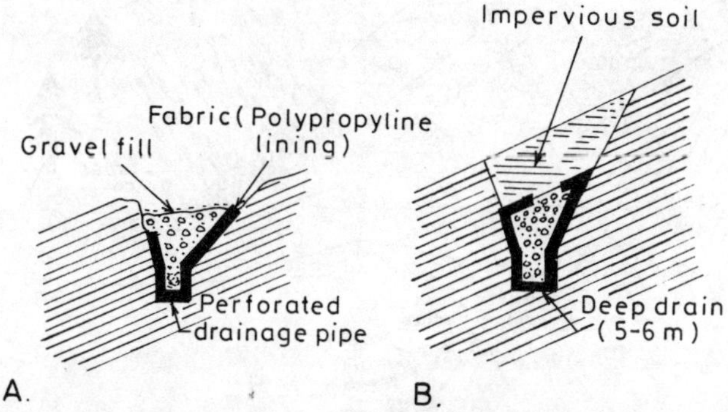


**Figure 9.10.** Drains constructed to remove water from the landslide-threatened/affected area should be lined by polypropylene sheets, and then filled with gravel in order to prevent it being clogged by garbage and debris falling from sides.

## VEGETAL TREATMENT OF DAMAGED SLOPES

To arrest erosion, it is advisable to provide a protection of vegetation to the landslide-affected/threatened area. The damaged slope is first fenced, and then seeded with quick-growing local grass (Figure 9.11). If the soil is admixed with appropriate fertilizer or leguminous plants, the grass will grow faster and richer. Mixing grass seeds with a small proportion of rapidly growing, stemmed plants such as sorghum (*jwar*) will promote richer growth of grass, for the sorghum serves as a nurse crop. Inoculation of seeds with *Rhizobium* bacterial culture will considerably quicken the growth of vegetation on the bare slopes. The seeded soil should then be covered by a net of coir, jute or synthetic yarn with an opening of about 2.5–1.5 cm (Figure 9.11A). The net is secured to the ground by overlapping fringes 5–8 cm wide, which are pegged down with staples or anchors. The net prevents the breaking of the soil (from rain or wind) and thus allows unhampered growth of the grass. Coir and jute nets eventually disintegrate and contribute manure to the soil.

When climatic conditions are unfavourable, the soil is covered with chopped straw, shredded bark and wood pulp. This mixed material lessens the heat from the sun's rays, and permits retention of moisture within the soil. To prevent erosion from rain and wind, an asphalt emulsion can be sprayed on the seeded soil just before the rain. The asphalt mulch eventually disintegrates, and contributes fertilizer to the soil.

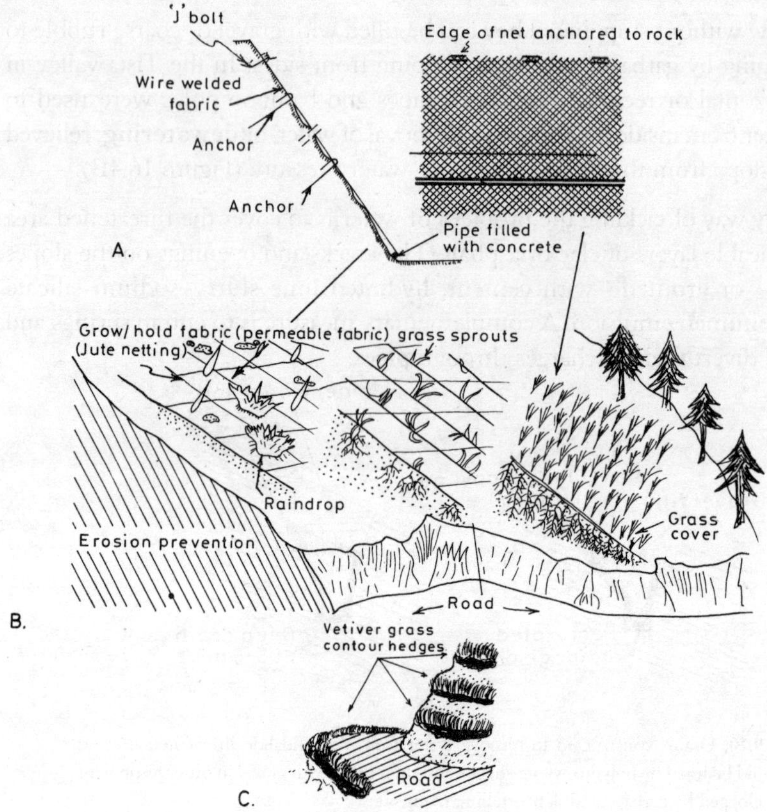


**Figure 9.11.** (A) The landslide-affected or threatened area is covered with a coir net, fixed to the ground by bolts or anchors. The net covers the soil seeded earlier with quick-growing grass.

(B) Coir net minimizes erosion by raindrops and wind, and allows the seeds to grow unhampered. Later, decomposition of the coir contributes to the manure in the soil.

(C) Vetiver grass has been found to make a very effective protective cover.

After partially stabilizing the slope with grass cover, the next step is planting shrubs and trees that have deeper roots and grow quickly. Not only do they dewater the soil through evapotranspiration, they also provide sorely needed shear strength to it by virtue of their deep and penetrating roots.

In recent years, Vetiver grass (*Vetiveria zizanoides*) (*khaskhas* in Hindi) has proved to be a versatile plant for protecting slopes (Figure 9.11C). It forms a dense hedge within 3–4 months of planting, and its dense and deep roots reach depths of two to three metres, greatly strengthening the slope material (Plate 5A, B).

## CHECKING MOVEMENT BY ENGINEERING STRUCTURES

A variety of **retaining walls** of concrete or masonry provide support to the threatened slope at its foot (Figure 9.12A). Low in height but heavy in weight, the **buttress walls** accomplish much the same objective (Figure 9.12B). These engineering structures not only prevent movements of the slope material, but also ensure protection against toe erosion along streams. The water that percolates through the affected rocks/soils poses a problem to these walls, as there is a build-up of pore-water pressure. It is, therefore, necessary to have a cushion of permeable gravel or coarse sand behind the walls, and weep holes in the structures themselves. The weep holes provide an outlet to the water accumulated behind the walls (Figure 9.12B). If possible, a monofilter or polyfilter or simply a polypropylene sheet with steel or wooden anchors (Figure 9.12A) should be placed before erecting the retaining and buttress walls. The filter prevents loss of solid material of the structure and stops water from entering it. Construction of drains up the slope, to check rain or surface water from affecting the structure, is a very desirable measure.

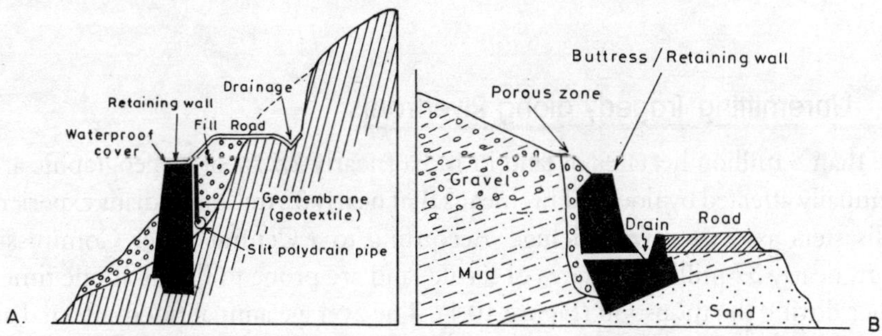


**Figure 9.12.** (A) A concrete retaining wall is keyed to the solid rocks of the slope in order to prevent rock falls and slides. Note that a drainage ditch at the top diverts surface water and there is a monofilter sheet behind the wall.

(B) Concrete buttress wall with a weep hole and a cushion of permeable gravel behind.

Quite effective in containing smaller landslides and slips are **gabion walls**, made up of steel-mesh cages or baskets filled with gravel and blocks of stones (Figure 4.12). **Crib walls** constructed with wooden poles or pre-cast concrete units provide equally effective support to the unstable slopes. However, if the slopes are threatened with creep movements, gabion and crib walls are not effective at all.

In order to reduce landslide hazards, it is imperative that the landslide type is identified on the basis of its activities and mechanisms. This calls for application of state-of-the-art techniques, which include remote sensing and collection of ground-truth data. The next step is to carry out measures for prevention and correction. The best strategy to control the occurrence of landslides is to stop the entry of water into the threatened or affected area as effectively as possible. The damaged slope and its soil need to be put under the protective cover of grasses, shrubs and trees that have deep roots. One must realize that a single remedial measure or two will not work in preventing landslides. It is always a package of measures that need to be taken to impart stability to the slope.

# 10

# Lessening the Ravages of Floods

## Unremitting Tragedy along Riverfronts

More than 8 million hectares of land in India (nearly 2.73% of its geographic area) are annually affected by floods. This means that nearly 32 million Indians experience the disasters associated with deluge. According to a Central Water Commission Report, nearly 37 million hectares of fertile land are prone to floods at one time or another during the monsoon (Figure 10.1). The average annual flood-affected area is about 7.4 m ha. This includes 3.8 m ha of *prized land*—crop fields that support about 32 million people. The Indo-Gangetic Plain, where more than 400 million people live in dense concentrations, bears the maximum impact of the floods (>60%).

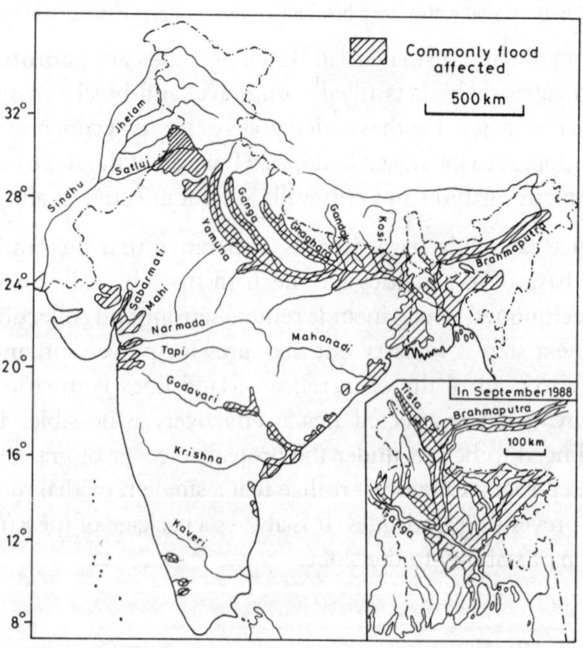


**Figure 10.1.** Flood-prone areas in India. The depicted areas are ravaged by deluge one time or another during the rainy season. *Inset* shows areas affected by floods in September 1998. (After M.C. Gupta et al, 2001).

The maximum flood-affected area in India varied between 1.46 m ha in 1965 and 17.5 m ha in 1978, the temporal variation and flood frequency showing no systematic trends (Kale, 1999). However, the population affected by the deluge has increased considerably—from 3.6 million in 1965 to 7 million in 1978—largely because population has been growing rapidly in the last 20–40 years (Sharma, 2000; Sharma, 2001). Since there are no regional trends in rainfall (Thapliyal and Kulshrestha, 1993), the increase in disasters and distress from flood is attributable to the explosive growth of population, and an excess of development that impedes the free flow of floodwaters.

The **recurrence interval** of severe floods is 2–3 years in the Indo-Gangetic Plain. In the Kosi–Gandaki floodplains (Figure 10.1), five to eleven floods occur every year (Kale, 1999). The situation in Bangladesh is much worse. In 1998, nearly two-thirds of the country's geographic area (8.28 m ha) was inundated (Figure 10.1, *inset*), taking a toll of 1,620 human lives (S. Goswami, 1998).

Engineering measures used to tame and control floods in the Himalayan rivers have failed. This is evident from the progressive increase in flooding in the Ganga–Brahmaputra floodplains. In 1998, floods occured in four frightening waves, deluging 38,200 km$^2$ (nearly 48.65%) of Assam and affecting the 12.5 million people of that region (Goswami, 1998). About 2.6 m ha of Bangladesh were inundated during floods. Analysis of sedimentary records of floods in the major rivers of central India and the Deccan plateau indicates that the magnitude and frequency of large floods increased progressively post-1940 (Kale, 1998; 1999).

**Table 10.1.** Devastating floods in the period 1980–96 (Gupta et al, 2001).

| Duration | Area affected | Synoptic systems |
|---|---|---|
| 17–23 July 1980 | East Uttar Pradesh | Low pressure area |
| 7–27 August 1980 | Central & southwest Uttar Pradesh | Land depression |
| 4–10 September 1980 | Southeast Uttar Pradesh | Land depression |
| 18–24 September 1980 | South Orissa | Deep depression |
| 18–24 September 1980 | Andhra Pradesh | Deep depression |
| 18–24 September 1980 | Central Uttar Pradesh | Deep depression |
| 9–15 July 1981 | Gujarat | Low pressure area |
| 16–22 July 1981 | Rajasthan | Low pressure area |
| 9–29 July 1981 | Uttar Pradesh | Low pressure area |
| 6–12 August 1981 | East Uttar Pradesh | Cyclonic storm |
| 3–9 September 1981 | East Uttar Pradesh | Land depression |
| 19–25 August 1982 | East Madhya Pradesh | Pronounced low pressure |
| 28–31 August 1982 | North Orissa | Depression |
| 30 Aug.–3 Sept. 1982 | Uttar Pradesh | Land depression |
| 20–23 June 1983 | Gujarat | Land depression |
| 11–17 August 1983 | West Maharashtra | Trough off the Maharashtra coast |
| 18–31 August 1983 | Northeast Uttar Pradesh | Low pressure area |

| Duration | Area affected | Synoptic systems |
|---|---|---|
| 15–19 September 1983 | Marathwada | Low pressure area west Maharashtra |
| 8–14 September 1983 | Southeast Uttar Pradesh | Land depression |
| 21–27 June 1984 | West Bengal | Pronounced low pressure |
| 28 June–11 July 1984 | Uttar Pradesh | Low pressure area |
| 23 Aug.–5 Sept. 1985 | Bihar | Low pressure area |
| 12–25 September 1985 | East Uttar Pradesh | Land depression |
| 11–18 September 1985 | Bihar | Pronounced low pressure |
| 17–23 July 1986 | Bihar | Land depression |
| 7–20 August 1986 | North Andhra Pradesh | Deep depression |
| 23–29 July 1987 | Bihar | Low pressure area |
| 30 July–20 Aug. 1987 | North Bengal | Cyclonic circulation |
| 3–16 September 1987 | Bihar plateau | Pronounced low pressure |
| 15–19 July 1988 | North Gujarat | Cyclonic storm |
| 25 Aug.–7 Sept. 1988 | North Gujarat | Cyclonic storm |
| 21–28 September 1988 | Panjab | Low pressure area |
| 19–26 July 1989 | Coastal Andhra Pradesh | Depression |
| 13–26 July 1989 | Western Maharasthra | Depression |
| 1–6 July 1990 | Northeast Rajasthan | Low pressure area |
| 16–29 August 1990 | Vidarbha | Depression |
| 23–29 August 1990 | Gujarat | Depression |
| 25–31 July 1991 | Vidarbha | Deep depression |
| 5–13 September 1991 | North Bengal | Cyclonic circulation |
| 16–22 July 1992 | Gujarat | Cyclonic circulation |
| 10–16 September 1992 | Jammu and Kashmir | Cyclonic circulation |
| 10–16 September 1992 | East Uttar Pradesh | Pronounced low pressure |
| 1–14 July 1993 | Gujarat | Cyclonic circulation |
| 8–14 July 1993 | Panjab | Cyclonic circulation |
| 15–20 July 1993 | Bihar plateau | Pronounced low pressure |
| 15–21 July 1993 | North Bengal | Pronounced low pressure |
| 9–15 September 1993 | Uttar Pradesh | Pronounced low pressure |
| 14–27 July 1994 | Gujarat | Low pressure area |
| 1–7 September 1994 | Orissa | Low pressure area |
| 1–7 September 1994 | Vidarbha | Pronounced low pressure |
| 17–23 August 1995 | Bihar | Low pressure area |
| 5–15 September 1995 | Haryana | Low pressure area |
| 20–26 June 1996 | Rajasthan | Low pressure area |
| 25 July–7 Aug. 1996 | Bihar | Low pressure area |

**Table 10.2.** Maximum flood discharge of some rivers when in spate (Kale, 1999).

| River | Highest Discharge ($m^3$/second) |
|---|---|
| Brahmaputra | 72,700 |
| Tista | 27,500 |
| Kosi | 25,800 |
| Damodar | 18,400 |
| Mahanadi | 46,000 |
| Narmada | 65,100 |
| Sabarmati | 10,900 |
| Sindhu | 34,000 |

## Causes of Recurrent Floods

There are three major causes of recurrent floods in the floodplains of India.

(i) When the magnitude of water flow exceeds the carrying capacity of a river channel within its banks, it results in water spreading into the floodway—the natural passage of the highest discharge. Heavy, prolonged rainfall and cloudbursts are the primary reasons for excess discharge. Excessive melting of snow and ice, accompanied by heavy rains, is also responsible for high discharge in Himalaya-born rivers. However, this development takes place rarely. More frequent causes are panic releases from reservoirs with excess water impounded behind their dams.

(ii) The bursting of dams—natural and artificial—have caused disastrous floods. In 1961, the collapse of the Khadakvasla and Panshet dams in the Mula river devastated a large part of Pune. The failure of the dam across the Machhu river in Saurashtra brought on a catastrophic deluge in 1979, that killed 1,419 inhabitants in the township of Morvi. On 10 June 2000, the bursting of the 130 m high debris dam (that had formed earlier as a result of landslides across the Palang Zambo river) washed away 20 big bridges, killed 30 people and rendered 50,000 people homeless in the Brahmaputra valley in Arunachal Pradesh. Likewise, the flood in 1994, when the glacial lake in Lugge Tso burst (Plate 13), caused extensive damage including the destruction of the Namche power plant close to the Phochu river in Bhutan (Mool, 2001). The 1994 bursting of the glacial lake in the Pho Chu river, in the Lunana area of Bhutan, is a classic case of hazard from ice melting (Joshi, 2001).

(iii) The carrying capacity—the water-holding capacity—of river channels diminishes appreciably due to the accumulation of sediments within their banks. Obstructions caused by the construction of buildings, factories, bridges, etc in the ways of floodwaters, considerably impede the free flow of rivers in spate (Figure 10.2A). Bridges with closely spaced piers or with protruding abutments restrict the passage of flood discharge. A floodway conveys its highest discharge once in a hundred years of floods, with one percent chance of recurrence, when the water-surface elevation does not rise to more than a foot above the floodway at any point. In other words, a **floodway** is the part of the valley which experiences one-foot inundation by floodwater at least once in a hundred-year period. Floodways are waterways that ensure relief from floods—if they are obstructed, there will be little relief from risks. Occupation and reclamation of floodways have led to tragic consequences whenever floodwaters rise above the safe level.

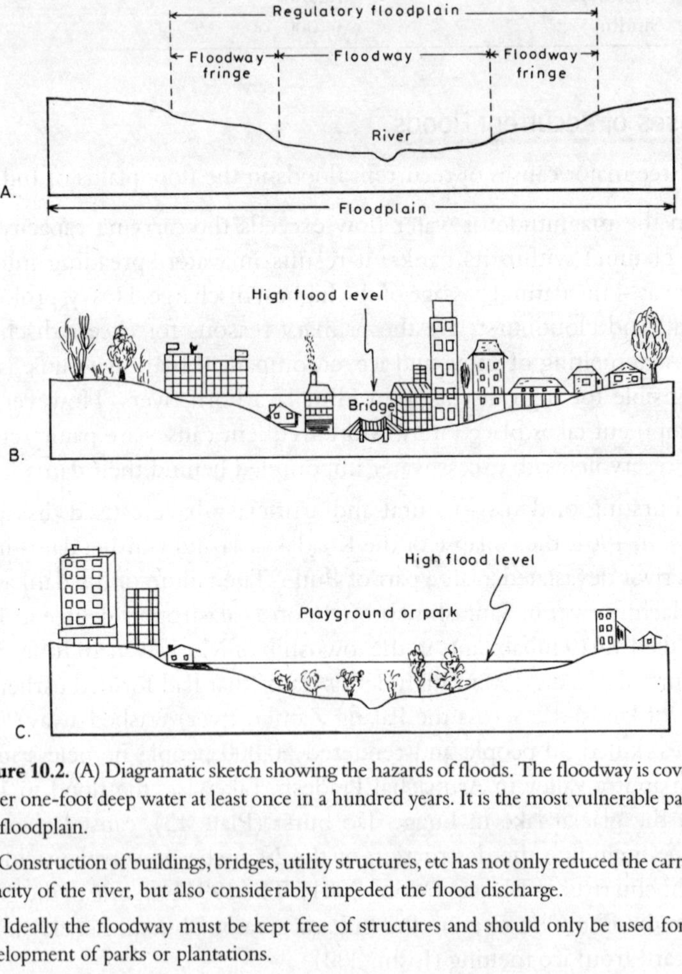


**Figure 10.2.** (A) Diagramatic sketch showing the hazards of floods. The floodway is covered under one-foot deep water at least once in a hundred years. It is the most vulnerable part of the floodplain.

(B) Construction of buildings, bridges, utility structures, etc has not only reduced the carrying capacity of the river, but also considerably impeded the flood discharge.

(C) Ideally the floodway must be kept free of structures and should only be used for the development of parks or plantations.

## EXCESS RAINWATER RUNOFF

The deprivation of vegetal cover in the catchments of practically all the rivers in the Indian subcontinent has resulted in accelerated erosion and excess surface runoff of rainwater. Where the rate of rainfall is higher than the rate of infiltration (into the ground) there is greater overland flow—so much so that a large proportion of rainwater flows on the ground surface as runoff, contributing to the making of floods (Figure 10.3).

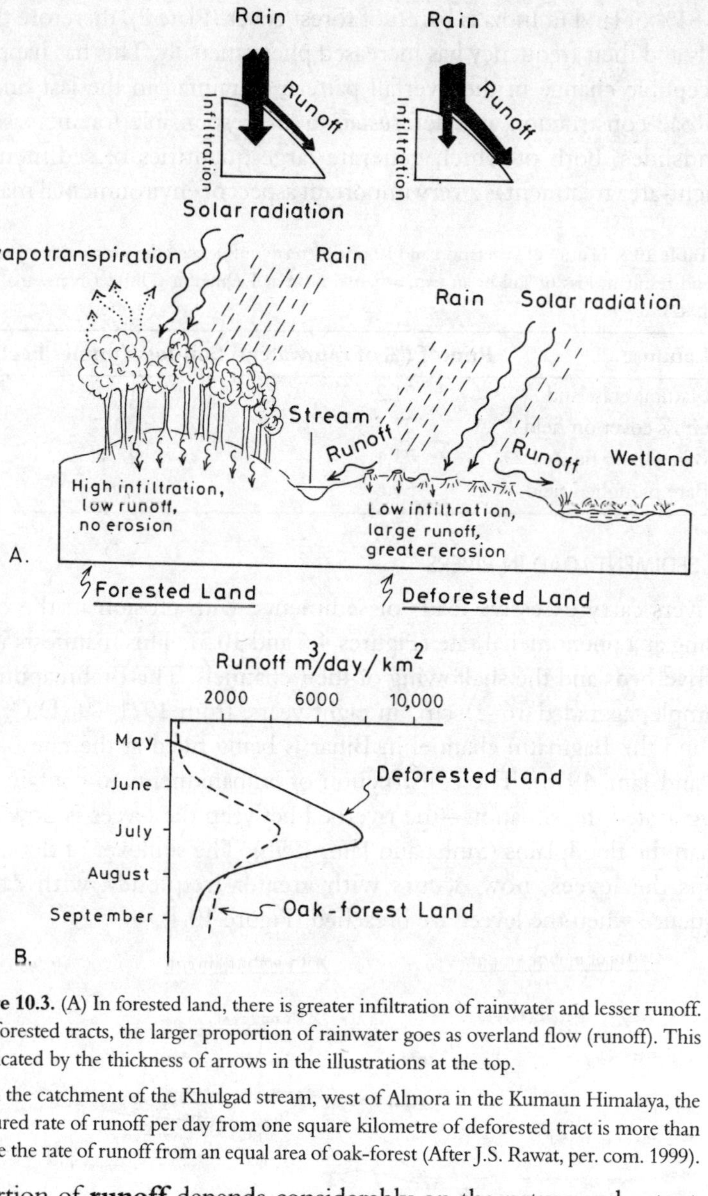


**Figure 10.3.** (A) In forested land, there is greater infiltration of rainwater and lesser runoff. In deforested tracts, the larger proportion of rainwater goes as overland flow (runoff). This is indicated by the thickness of arrows in the illustrations at the top.

(B) In the catchment of the Khulgad stream, west of Almora in the Kumaun Himalaya, the measured rate of runoff per day from one square kilometre of deforested tract is more than double the rate of runoff from an equal area of oak-forest (After J.S. Rawat, per. com. 1999).

The proportion of **runoff** depends considerably on the nature and extent of the vegetation growing in the catchment. Vegetation generates plentiful litter that permits greater **infiltration** in a variety of ways. Converted to humus, the vegetal matter increases soil permeability, and consequently rainwater infiltration. Clearing, burning and overgrazing of forests in catchments have greatly increased peak discharges of streams and rivers. In an experimental farm at Dehradun, the clearing of a forest of *saal* trees resulted in an increase in the volume of runoff by 15%, and of the peak discharge of streams by 72% (Dhruva Narayana et al, 1997). Table 10.3 explains how the nature of vegetation and human activities influence the proportion of overland flow and the extent of resultant soil loss.

Nearly 81% of land in India is bereft of forest cover (Plate 2), therefore the severity of floods and their frequency has increased phenomenally. This has happened with no perceptible change in the overfall pattern of rainfall in the last one hundred years. Road construction and deforestation are responsible for increased erosion and landslides, both of which generate large quantities of sediments. Hence catchment-area treatment is a very important aspect of environmental management.

**Table 10.3.** Nature of vegetation and land use greatly influenced the proportion of runoff and resultant loss of soil in an experimental farm at Dehradun (Dhruva Narayana et al, 1997).

| Landuse | Runoff (% of rainwater) | Soil-loss (tonnes/hectare) |
|---|---|---|
| Natural grassland | 21.2 | 1 |
| Grass cover on field | 27.1 | 2.1 |
| Bare fallow field | 71.1 | 42.4 |
| Bare ploughed field | 59.6 | 156 |

## EXCESS SEDIMENT LOAD IN RIVERS

Most rivers carry excessive loads of sediments, with erosion in the catchments increasing at a phenomenal rate (Figures 4.6 and 10.3). This manifests itself in the rise of riverbeds and the shallowing of their channels. The Brahmaputra channel, for example, aggraded 16–21 cm in eight years, from 1971–79 (D.C. Goswami, 1998), and the Bagmathi channel in Bihar is being filled at the rate of 9.3 cm/yr (Sinha and Jain, 1998). The construction of embankments to contain floods has only aggravated the situation—the riverbed between the levees is now at a higher level than the floodplains (Sinha and Jain, 1998). The spillover of floodwater, as it overtops the levees, now occurs with greater frequency, with catastrophic consequence when the levees are breached (Figure 10.4).

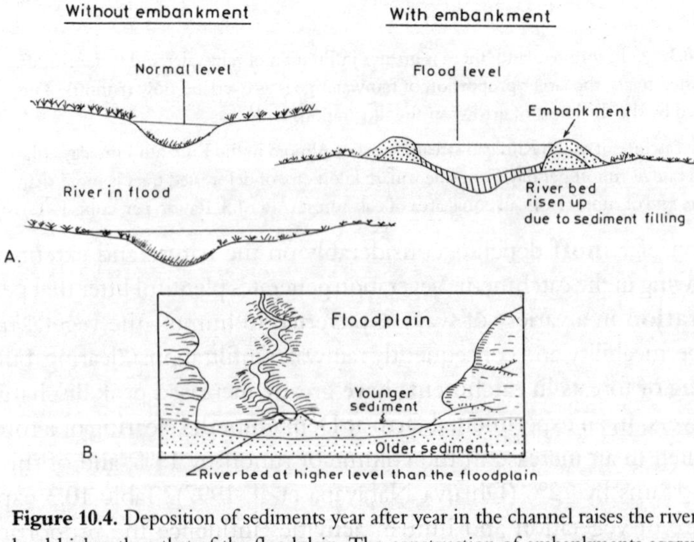


**Figure 10.4.** Deposition of sediments year after year in the channel raises the riverbed to a level higher than that of the floodplain. The construction of embankments aggravates the situation. Overtopping of, or breach in, the levees brings disastrous floods.

Another consequence of sediment deposition in river channels is the progressive reduction of the carrying capacity of channels. Even the formation of pointbars at riverbends, due to accumulation of sediments, contributes to the slackening of the flow. This results in the spread of floodwater onto the floodplains. In the case of the Gandak and Kosi rivers in Bihar, the width–depth ratio in the former is 133:117 and in the latter is upto 78, which indicates copious accommodation for water and sediments, as borne out by braiding and multitudes of channel bars (Sinha and Jain, 1998). The best method to control overbank flooding is channel gradient. However, large volumes of sediment within these channels indicate that their carrying capacity has drastically reduced. This explains the wild nature of these rivers, often described as the 'Sorrow of Bihar'. In Pakistan, the Sindhu frequently spills over its embankments, in the province of Sindh, for the same reason (Masud, 2001).

Deposition of large volumes of sediments resulted in the formation of large bars and islands in the 8–10-km wide channel of the Brahmaputra river in Assam, between Guwahati and Pasighat (Plate 14, also Plate 4B). The braided river seems to be filled with sediments. This explains why Assam is repeatedly ravaged by uncontrollable floods. As mentioned early in the chapter, in 1998, 48.65% of Assam was inundated by floodwater.

## Coping With Flood Hazards

### APPROACH

A typical flood management programme embodies, at first, the preparation of a **flood-zoning map**. Such maps should: (i) demarcate areas that are recurrently inundated, (ii) illustrate areas where construction cannot be allowed, and (iii) point to areas where recharge of groundwater takes place.

The programme should be further strengthened by a system of monitoring flood-discharge at critical places, and giving timely warning to people in the threatened areas. The information required for such monitoring includes: (i) the amount of rainfall in the catchment area and in the river basin, (ii) the volume of discharge and the velocity of floodwater at critical points, (iii) the likely height the floodwater would reach, and (iv) the duration of the deluge in the threatened area.

Application of remote-sensing techniques has proved extremely useful in both flood-zone mapping, and monitoring river discharge and the spread of the floodwater. These images can form the basis of detailed maps on a scale 1:25,000 or 1:50,000, or even larger scale topographic maps.

### WATERSHED TREATMENT: A SUSTAINABLE MEASURE

An ecologically sound management philosophy requires regional watershed planning, which is by far the most sustainable way of reducing flood hazards. This also entails action, and decision-making, at local levels. The plan involves measures that will reduce runoff in the catchment itself. **Watershed treatment** has been dealt with in Chapter 4 (Figure 4.11). In order to reduce the runoff proportion of rainwater, it is desirable to make the soils permeable and induce percolation of the rainwater down to underground aquifers. This can be achieved by cultivating vegetal cover that generates abundant litter, and by digging trenches, pits and wells to induce infiltration. Development of spring sanctuaries complements this endeavour in a very holistic way.

### IMPOUNDMENTS IN CATCHMENTS

Storing rainwater—in **detention basins** and reservoirs in upper reaches, and in wetland lakes, wetlands and paddy fields in the floodplains themselves—is an effective method of abating flood hazards. It is imperative that reservoirs developed for flood control are kept only partially filled, if not quite empty. Construction of a multiplicity of smaller dams (including check dams) in as many tributary streams as possible would help reduce sudden, excessive discharge in the main river. Operating dams on large tributaries, with relation to flood warning, needs to be done effectively.

### BUILDING EMBANKMENTS ALONG RIVERS

Conventional methods being followed to control floods involve building **embankments** parallel to the floodways. This engineering structure prevents or reduces the spread of excessive discharge onto the floodplain (Figure 10.5). However, this structural approach is no longer the primary focus of floodplain management in advanced countries like the U.S.A. In India, this embankment strategy has not succeeded in containing flood-fury in the Indo-Gangetic Plain. An aggregate of 12,265 km of embankments, with 21,810 km of drainage channels, were constructed between 1952–80, but in vain. Compared to nearly 6 m ha of land affected by floods in 1953, approximately 11.8 m ha were under floodwaters in 1980 (Patnaik, 1978). Despite being bunded by 246 km of embankments, the Kosi remains an untamed river. In Assam, embankments built to the length of 3,642 km, with 599 km of drainage channels, did not contain floods in the Brahmaputra river (S. Goswami, 1998), which continue to wreak havoc year after year.

It is obvious that embankments do not provide relief from flood hazards. In all the measures that tame a river, it is advisable to allow the river to move freely in its **meander belt**. Therefore, embankments must be built quite beyond the **flood-fringe** (Figure 10.5).

### CHANNELIZING FLOODWATER

If excessive discharge can be distributed through a network of canals and interlinked rivulets and streams, the hazard of flood would considerably diminish. Each of the channels and streams could serve as a temporary storage for short durations. If one considers the extent of the network of canals and interlinked streams, one would realize that the total aggregate storage for floodwater is considerable. An extension of channeling floodwater is the inter-basin transfer of waters.

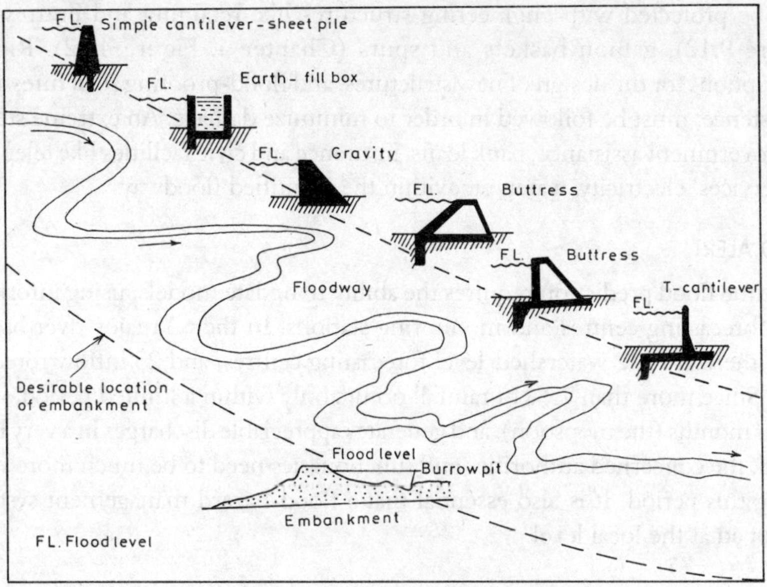


**Figure 10.5.** Embankments to contain flood-fury must be built quite beyond the flood-fringe so that the river in spate is free to swing in its meander belt. Cross sections of various models of embankment are shown along the left-bank side.

A complementary measure is to spread the floodwater onto the paddy fields and natural depressions and allow it to percolate into the ground (Figure 5.16). If half of the nearly six million hectares of paddy fields were to be inundated with 20-cm deep water, nearly five million hectare-metres of floodwater could be prevented from bringing on a deluge.

The **channelization** of floodwater has its own problems. The deepening, widening and straightening of a river course increases its ability of conveyance, as the discharge becomes quicker and larger. This quickens the pace of channel-erosion. The swifter flow adversely affects aquatic life accustomed to a quieter pace and flow. This may also result in the elimination of breeding areas. Channelizing may minimize floods upstream, but there would be severe problems downstream due to increase in flow. Channelization is therefore preferred when the river is to be used for navigation. In such cases, bank protection and velocity reduction are realized by constructing dams or navigational locks.

## FLOODWAY REGULATIONS

Flood-zoning exercises include the identification and precise demarcation of floodways. The floodway (Figure 10.2; Plates 14 and 15) is the critical belt of the floodplain that is prone to inundation—one foot deep at least once in a hundred years. It is this belt which must be kept free of all obstructing structures like buildings, walls, bridges, etc. This needs to be adopted and implemented as common floodway regulation measures. There must be no construction of any kind, either for urbanization or industrialization, within the floodway. The flood-fringe (Figure 10.2) must be protected with engineering structures like retaining and buttress walls (Figure 9.12), gabion baskets and spurs (Chapter 4, Figure 4.12). Rigorous prescriptions for the design of new structures, and flood-proofing structures already in existence, must be followed in order to minimize damage. An extreme step is to stop government assistance, bank loans, insurance and civic facilities like telephones, bus services, electricity, water, etc within the identified floodway.

## FLOOD ALERT

Real-time flood prediction requires the ability to update models, using information from forecasting centres and monitoring stations. In the 62 major river basins in India, there are 132 watershed-level forecasting centres, and 25 inflow forecasting units. Since more than 75% of rainfall occurs only within a limited period of three to four months (the monsoon), and generates appreciable discharges in a very limited period, the concerned authorities and functionaries need to be much more vigilant during this period. It is also essential that a flood-hazard management system be developed at the local level.

# 11

# Tempering Tempests and Coping with Coastal Cyclones

## Hazard-prone Belts

Along the Indian coast, a belt 24-km wide has been identified as prone to recurrent coastal hazards from tropical cyclones (Plate 16) and storm surges (Figure 11.1). This belt stretches from coastal Bangladesh to Thanjavur district in Tamilnadu (along the east coast), and from Bhavnagar district to Junagarh district in Gujarat (along the west coast). The impact of coastal hazards is not restricted to this narrow belt alone—it stretches inland by as much as a hundred kilometres when fierce winds sweep across the seaboard.

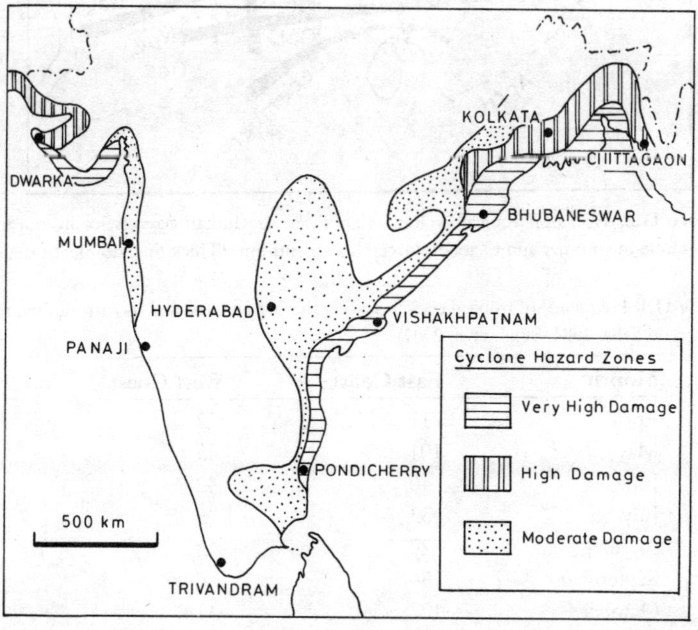


**Figure 11.1.** Belts identified as prone to coastal hazards from cyclones, tempests and storm surges (Based on the *Natural Hazard Map*, National Atlas & Thematic Maps Organization, Kolkata, in Gupta et al, 2001).

## Severity of Cyclones

### FREQUENCY OF TROPICAL CYCLONES

A little more than 7% of tropical cyclones the world over originate in the Bay of Bengal (Figure 11.2). On average, five to six cyclones originate in the Bay of Bengal and Arabian Sea every year. Cyclones and attendant storm surges occur mostly in summer (April–June) and autumn (October–December). An analysis of data over the past 122 years (1877–1998) shows that there has been a two-fold increase in the frequency of intense tropical cyclones, primarily during the months of May and November, in the Bay of Bengal. Conversely, there is a decrease in frequency in the transitional monsoon months of June and September (Singh et al, 2001). Cyclones visit the east coast four to six times more frequently than the west coast (Table 11.1). Along the Bangladesh and Andhra coasts, they occur once almost every two years. The life span of a common tropical cyclone is four to five days.

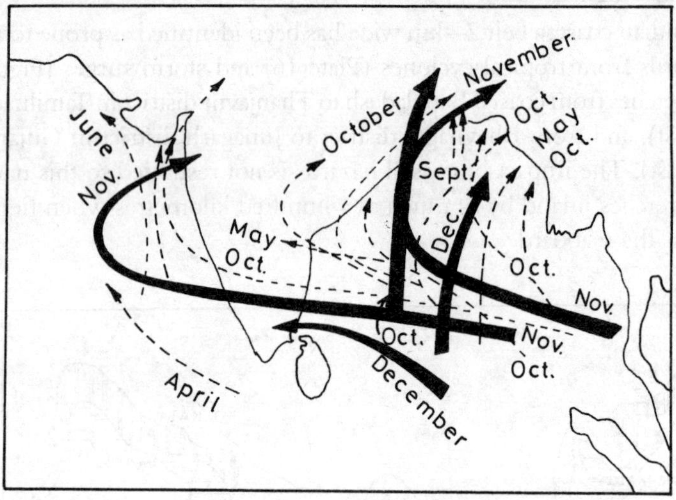


**Figure 11.2.** Tracks of tropical cyclones that strike the Indian coast twice in a year—April–June in summer and October–December in autumn. (Thick line: dominant trend)

**Table 11.1.** Frequency of tropical cyclones along the Indian coast and its surrounding seas (Based on Saha, 1984; Singh et al, 2001).

| Month | East Coast | West Coast |
|---|---|---|
| April | 11 | 2 |
| May | 101 | 43 |
| June | 40 | 42 |
| July | 33 | 3 |
| August | 27 | 2 |
| September | 56 | 6 |
| October | 127 | 35 |
| November | 177 | 35 |
| December | 22 | 4 |
| Total | 594 | 172 |

## LOSS OF LIFE

In November 1970, a storm that killed 200,000 people and destroyed more than 200,000 houses in Bangladesh, had wind speeds of 224 km/hr close to the centre, very heavy rainfall, and storm surges a few metres in height. 800,000 heads of cattle lost their lives and more than 80% of the standing paddy crop was destroyed. Between April 25–30, once again in Bangladesh, a cyclone with wind speeds of 235 km/hr took a toll of 139,000 people. In 1999, the October 17–29 cyclone hit the Orissa coast at speeds of 260–300 km/hr—killing over 20,000 people, severely damaging 2 m ha of farmland, uprooting 200,000 trees, and washing away 5,200 km of roads. Farmlands were ravaged by persistent salinity from tidal waves. Storm surges rose 8 m high, while blinding rains lashed down for 48 hours (50 cm) and submerged the coastal area in 3–4 metres of water.

**Table. 11.1.** Some severe cyclones in India (Gupta et al, 2001).

| Location | Date | Damages |
|---|---|---|
| West Bengal | October 1847 | 75,000 people and 6,000 cattle killed. Damage to property and communication systems. |
| West Bengal | October 1874 | 80,000 people killed. Heavy loss to property and communication disrupted |
| Andhra Pradesh | November 1946 | 750 people and 30,000 cattle killed. Damage to property and roads. |
| Tamilnadu | December 1972 | 80 people and 150 cattle killed. Damage to property, including communication. |
| West Bengal | September 1976 | 10 people and 40,000 cattle killed. Damage to property, including communication. |
| Andhra Pradesh | November 1977 | 8,547 people and 40,000 cattle killed. Communication disrupted, heavy loss to property. |
| Tamilnadu | May 1979 | 700 people and 300,000 cattle killed. Communication disrupted. |
| Orissa | September 1985 | 84 people and 2,600 cattle killed. 4 m ha of land damaged. |
| Andhra Pradesh | November 1987 | 50 people and 25,800 cattle killed. 8,400 houses destroyed, roads and other communication disrupted |
| Orissa | June 1989 | 61 people and 27,000 cattle killed. 145,000 houses damaged, communication disrupted. |
| Andhra Pradesh | May 1990 | 928 human lives lost. 14,000 houses damaged. |
| Tamilnadu | November 1991 | 185 people and 540 cattle killed. Extensive road damage. |
| West Bengal | April 1993 | Over 100 casualties. Communication systems including roads disrupted and damaged. |

| Location | Date | Damages |
|---|---|---|
| West Bengal | November 1994 | More than a thousand houses damaged in 26 villages. Damage to lakes and fisheries, all communication disrupted. |
| Andhra Pradesh | October 1996 | 1,057 casualties. 647,000 houses damaged, road network completely damaged. |
| Gujarat | June 1998 | 1,261 casualties. 25,700 houses damaged. |
| Orissa | October 1999 | 10,086 casualties. 216,000 houses damaged. |

### IMPACT ON THE ENVIRONMENT

Cyclonic storms have appreciably reshaped landforms along the Andhra–Orissa coast, particularly in the deltaic tracts (Plate 17) of the Mahanadi, Godavari and Krishna rivers. Beach sands were exposed to erosion, giving rise to lobate sandbars between Machhilipatnam and Machugunta. Not only croplands, large tracts of forests were also covered by sand, and much larger areas became waterlogged. In some areas, accumulation of sand behind impediments (such as a lighthouse) gave rise to sandbars, some as high as 40–90 m above ground level, as witnessed at Pudimadka and Bhimla (Prudhvi Raju and Krishna Bhagwan, 1984). The Orissa supercyclone caused great change in the sedimentation pattern along the coastal belt—there was rapid siltation in the delta region, and in estuaries, lagoons and ports; but depletion in the offshore belt (Nayak et al, 2001).

Not only are surface water bodies contaminated by saltwater from the sea, the soils of the cultivated fields also become salt-encrusted. This happened in the Paradip coastal belt during the 1999 supercyclone. Earlier in 1977, more than 1.46 m ha of cropland in Andhra Pradesh were severely affected by this phenomenon.

## Genesis of Cyclones and Storm Surges

Spirally moving fierce winds forming in the Bay of Bengal have diameters of a few hundred kilometres. In the central area is a zone of a few kilometres (usually less than a hundred kilometres across) where the winds are calm. This is called the **eye of the storm**. Counter-clockwise rotating winds attain a speed of more than 200 km/hour along the periphery of the eye, but remain quiet at the centre—the eye. Cyclonic storms advance towards low pressure areas on land with speeds of 10–20 km/hr (Plate 16). Lasting on average for five to six days, cyclonic storms lose their fury within 24 hours of landfall. In some cases, the storms re-enter the sea, revive, and return with greater severity. The Orissa cyclone of 17 October 1999 moved very fast over the sea, but slowed down just before landfall. It then intensified into a very severe cyclone. The second cyclone on 29 October 1999 turned into a supercyclone just before hitting the land. It stagnated over the coastal region for two days and finally zigzagged back to the sea (Singh et al, 2001).

Generally, when seawater in this tropical belt (5–15°N) heats up to 27°C or above, and is aided by convective cloud areas and suitable wind fields, it gives rise to low pressure areas over the southern Bay of Bengal and southeastern Arabian Sea (Figure

11.2 and Plate 15). The pressure starts falling and drops by 20–40 millibars in a few days. Remaining stationary or moving slowly for three to four days, the low pressure area draws energy from the surface of the sea. Air from the surrounding areas rushes into the low pressure centre. The rushing wind curves and accelerates, and can attain speeds that may exceed even 200 km/hr. Cloud bands start spiralling around the centre, bringing squalls in their trails. The pressure at the centre drops continuously and the converging warmed-up air may rise as high as 12,000 m or more above sea level. The result is an anti-clockwise spiralling wind (in the northern hemisphere) that may attain speeds of 100–120 km/hr often, and over 200 km/hr sometimes. This is the **cyclonic storm**, known as a **hurricane** in the Atlantic Ocean and **typhoon** in the Pacific Ocean. Exceptionally heavy rains exceeding 25–75 cm, accompanied by strong winds and storm surges (during the short passage of a cyclone) can cause flood havoc. During the cyclone of late-November 1984, the Andhra coast received 110 cm of rainfall.

The passage of cyclones causes a sudden reduction of atmospheric pressure over the coastal area—as much as 60–140 millibars or more. This development generates **storm surges**, which may rise as high as 14 m. On 7 October 1737, the township of Kolkata was hit by 12-m high storm surges, killing nearly 300,000 people. In 1864, storm surges nearly 12-m high struck Contai (northeast of Digha, in Orissa) and killed 50,000 people and 10,000 heads of cattle. During the 1999 Orissa cyclone, storm surges were seven metres high. The storm water went deep inland (15 km) through waterways and tidal inlets, destroying 90% of croplands and devastating 320,000 ha of land. Between 5–6 June 1998, the Kachchh coast (in Gujarat) was ravaged by 18–20-m high storm surges, which resulted from a cyclone that stormed inland with core wind speeds of 235 km/hr, along the Kandla coast.

## Strategies to Mitigate Cyclone Hazards

### STRENGTHENING WARNING SYSTEMS

The first step in hazard-management exercises is to draw up a **vulnerability map** after analyzing the situation. It is as necessary for managers to have a reasonable perception of the magnitude of an impending hazard, as there should be a willingness amongst the hazard-prone population to follow instructions in order to reduce the extent of the disaster. Timely warning will enable the threatened people to evacuate the region.

Developing an effective and efficient system to monitor the movements of cyclones through satellites, radar and aircraft is an essential part of a hazard-preparedness plan. Meteorologists can now forecast the occurrence of cyclones and delineate the path of its possible movement. Remote-sensing methods—including the application of Doppler S-band radar-tracking and analysis of satellite images—provide a very powerful and reliable tool to predict the trend, tract, and time of landfall of cyclonic storms. The forecast should be followed by timely information and public warnings so that the hazard-prone people can seek safety. Satellite-based surveillance in 1990 enabled the Andhra Pradesh authorities to evacuate 60,000 people in time. Thus, the death toll was limited to 1,000 people, in comparison with 10,000 deaths in the earlier cyclone of 1977 (Rao, 2000).

## PROTECTING COASTS: ENGINEERING MEASURES

Erosion along the coasts can be brought to a minimum using engineering measures. As already explained in Chapter 4 (Figure 4.13), shelter belts of plants (trees and shrubs with grasses) parallel to the coast and close to the shore, provide insurance against the fury of waves and storm surges. Creepers like *Ipomea biloba* can be planted along sand dunes to check the movement of sediments. Erection of palisade fencing helps initiate sand accumulation. Removal of coastal sand, to be used as construction material in these areas, must be controlled or stopped altogether.

Building of masonry or concrete structures like **seawalls**, **breakwaters** and **groynes** (Figures 11.3 and 11.4) reduce the intensity of waves pounding the coasts. Offshore breakwaters intercept waves and create protected areas. Groynes built normal to the coastline at intervals (Figure 11.4), help to arrest erosion as they build beaches by accumulating sands downstream of longshore currents. Such structures have proved fairly effective in arresting erosion off the coast of Kerala (Baba, 1979) and Orissa (Acharyya et al, 2000). **Gabion baskets** (Figure 11.5) and **pile-crib walls**, have successfully saved the coasts from wave erosion.

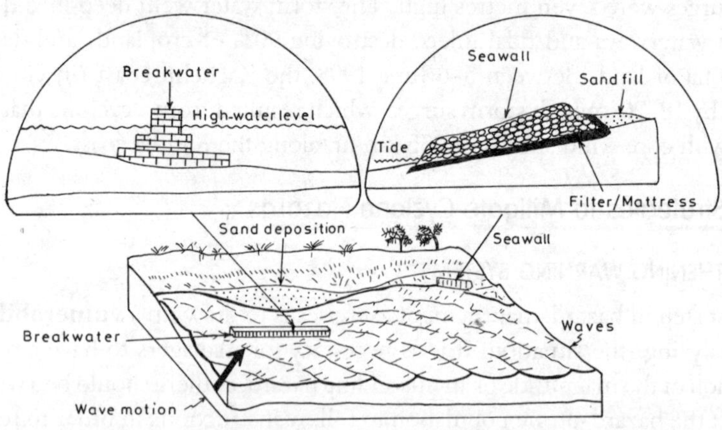


**Figure 11.3.** Offshore breakwaters and seawalls along the beach reduce the intensity of waves. The top two *insets* show the shape of the structures and positions of these protective walls in detail.

## CONSTRUCTING CYCLONE-RESISTANT HOUSES

Built on raised grounds or platforms so that they are above inundation, and anchored firmly to the ground at the corners, the circular-plan huts with steep roofs that fishermen of the Andhra coast use, seem to be quite resistant to the damage of cyclonic storms. A central anchor post, with metal corner straps to fix frame-members to the corner posts and beams, is the main feature. The thatched houses have: (i) internal partitions in order to increase their lateral stability, (ii) diagonal

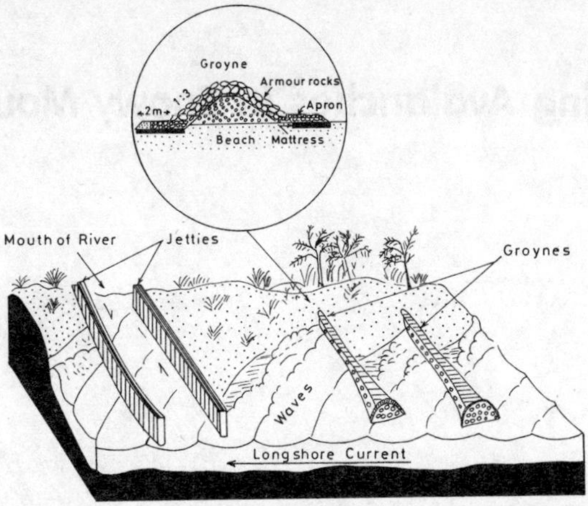


**Figure 11.4.** Groynes, built normal to the coastline, help accumulation of sand carried by longshore currents, thus contributing to beach nourishment. *Inset* shows the structure of a groyne in cross-section.

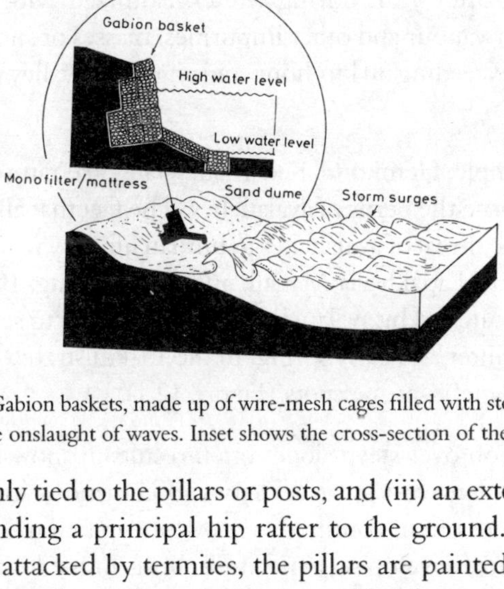


**Figure 11.5.** Gabion baskets, made up of wire-mesh cages filled with stones, protect sandy coasts from the onslaught of waves. Inset shows the cross-section of the structure.

inward braces firmly tied to the pillars or posts, and (iii) an external prop to the roof structure by extending a principal hip rafter to the ground. To save them from rotting and being attacked by termites, the pillars are painted thickly with asphalt and covered with polythene sheets at the ground level.

Masonry walls in houses should have bands at lintel level, and these bands must be anchored to the rafters and the walls. The roof should be firmly tied or securely bolted to the walls. The facades of the house should be bleak, with minimum apertures in the direction of prevailing winds. The shorter side of the building should face the storm.

Houses in such areas should be located on elevated places, and be sheltered by *Casuarina* hedges. It is however much more desirable that settlements should be far away from coasts that are prone to storms. The safer places would be behind the natural barriers of sand dunes, tree plantations and shelter belts, where the violence of high waves is reduced.

# 12

# Averting Avalanches in Snowy Mountains

## Hazards in the Himalaya

The celebrated Badarinath shrine in Garhwal withstands the threat of moving masses of snow and ice, year after year, during winter. Admixed with ice crystals, and permeated by air, water vapour and other impurities, masses of snow descend down steep slopes at speeds exceeding 30 km/hour, and engulf the valley of the headwaters of the Ganga.

Like the Badarinath temple, Hemkund, Kedarnath, Gangotri, Yamunotri, Trilokinath and Amarnath have borne the perils of avalanches. The fact that all these are centres of pilgrimage makes the threat of avalanches to human life very real. The geopolitical and strategic highway to Ladakh via Manali, and the Rohtang, Banihal and Zojila Pass sectors, are all threatened by avalanches that occur in six to seven waves, in the 20–24 days between winter and early spring. In the 13-km stretch of the Zojila Pass alone there are 37 avalanche-prone spots (Figure 12.1).

Rapid snow precipitation over steep slopes often results in snow hurtling down as a billowy mass of powder, with great velocity and enormous impact. This is the **snow avalanche**. Quite often, continuous snow precipitation over long periods can exert ample pressure, developing a metamorphosis in the lower layers of snow (Sharma, 2004). The lower layers could melt under this pressure and buckle under the overlying load—the net result being the complete sliding down of the overburden. Snow avalanches are often catastrophic in nature. The speed of avalanches varies between 30–100 km/hour, and pressure has been estimated at between 5–50 tonnes/$m^2$.

In Kashmir, the impact pressure is between 20–30 t/$m^2$ (Mohan Rao and Rangachari, 1983). Understandably, all that comes in the way of the avalanche is destroyed. Over 600 human lives were lost in the Jammu–Kashmir section of the Himalaya in 1978–79. In March 1979, in the Lahaul–Spiti region in Himachal Pradesh, 237 people were killed in an avalanche. The average annual death toll in the entire Himalayan region is about 30 persons.

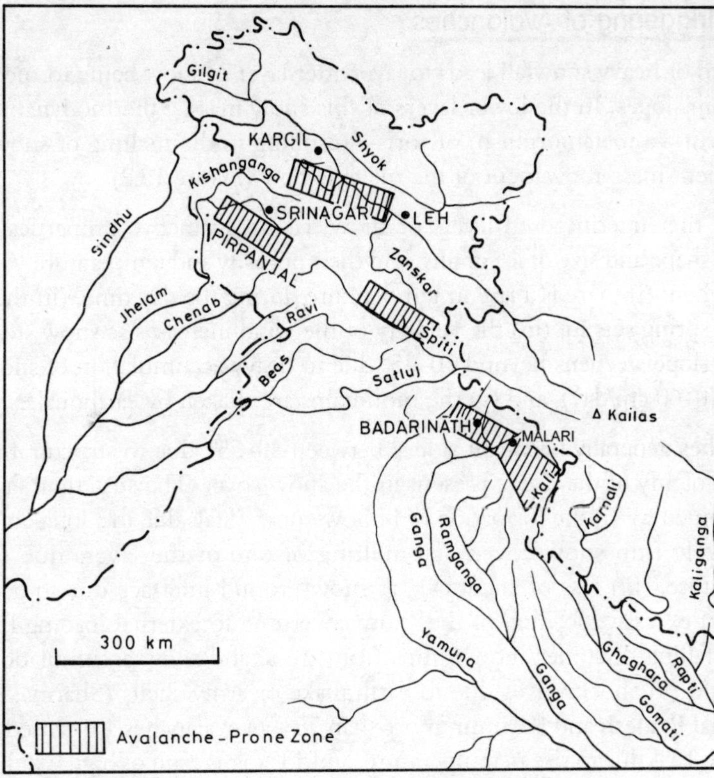


**Figure 12.1.** Avalanche-prone areas in the northwestern part of the Himalaya.

**Table 12.1.** Avalanche disasters in India (Gupta et al, 2001).

| State | Year | Damage |
|---|---|---|
| Himachal Pradesh | March 1978 | 30 people killed. Road and property damaged. |
| Lahaul and Spiti (H.P.) | March 1979 | 237 people killed. Communication lines disrupted. |
| Jammu and Kashmir | December 1982 | 100 people killed. Communication lines disrupted. |
| Ladakh (J & K) | December 1984 | 27 soldiers killed in an avalanche generated by artillery fire. |
| Jammu & Kashmir | March 1988 | 70 people killed, communication lines disrupted. |
| Himachal Pradesh | March 1991 | An avalanche takes place 4–5 times every year at Tinku, between January–March. The road was blocked for 40 days in 1991. |
| Himachal Pradesh | September 1995 | The avalanche caused the descent of huge chunks of debris, which later melted to create a flood. |

## Triggering of Avalanches

Repeated or heavy snowfall leads to layer after layer of snow being formed on high mountain slopes. In the lower layers of this snow mass, a thermodynamic change may occur—a metamorphism of sorts—resulting in the melting of snow and the consequent mass-movement of the melted snow (Figure 12.2).

Each of the superimposed layers of snow has its distinctive properties—specific density, shape and size of ice grains, and their porosity and temperature. Avalanches occur when: (i) there is a rise in temperature during the day time, (ii) the warmth of early spring sets in, (iii) the velocity of the wind increases as snow storms rage, (iv) the slope steepens beyond 30–45° due to heavy accumulation of snow (at the rate of 40–60 cm/day), and (v) the mountains are shaken by earthquakes.

Avalanches generally occur on slopes between 30–55°, due to structural failure or collapse of any of the layers present in the snow cover. The structural failure may be attributed to: (i) the formation of hollow cup-crystals due to a long cold spell in a relatively thin snow cover, (ii) melting of one of the layers due to rise in temperature, (iii) loss of anchorage at snow-ground interface due to geothermal heat, (iv) excessive loading of the snow cover due to external loading by human beings, falling of stones/snow lumps from trees, and snow accumulation due to wind, and (v) shock waves due to earthquake or army shells (Sharma, 2004). In Himachal Pradesh and Kashmir, more than 75% of avalanches are caused by heavy snowfall, 15% due to rise in temperature, and 13% as a result of high wind speed. In the Zojila and Banihal Pass sectors, wind velocity in excess of 15 km/hr is known to have triggered severe avalanches.

## Types of Avalanches

Avalanches occur primarily in two forms. When a mass of loose crystals of snow and ice—admixed with air, water and other impurities—is set in motion, part of it gets airborne as it drops down the precipitous slope, moving like a mass of loose powder. The ground component hurtles down the slope at speeds between 30–100 km/hr and flows down the valley like a river of snow. This is a **fresh snow avalanche**, also called direct action avalanche (Figure 12.2A and Plate 18A). It has very strong pressure and creates shock waves in the valley below. These avalanches generally occur during early- or mid-winter, or immediately after snow storms—initiated by the accumulation of fresh snow. They may contain part of the old accumulated snow in the form of hard or soft slabs.

When snow accumulates progressively for a longer period and forms a thick pile, the load on the lower layers increases to such an extent that the layers become cohesive or hard, made up of an aggregate of rounded crystals. Sudden shocks—generated by explosions, earthquakes, movements of heavy vehicles, and even skiers—cause a breaking within the hardened layers and propel the layers downslope. As it slides down the steep slope, the snow pile cracks up, breaking into slabs, blocks and fragments (Figure 12.2B and Plate 18B). This is called a **slab avalanche**. In the Kashmir Himalaya, the diurnal temperature variation is 6–10°C, enough to cause snow melt. This helps in triggering slab avalanches. In the upper parts of the Himalaya, during a long cold spell, the formation of a weak layer inside the snow pack and its fracturing due to external forces or vibration, can also result in an avalanche. These delayed and unpredictable avalanches have become the bane of strategic highways in Ladakh. The impact pressure of slab avalanches is 20–30 tonnes/$m^2$ in the Kargil–Leh sector (Mohan Rao and Rangachari, 1983).

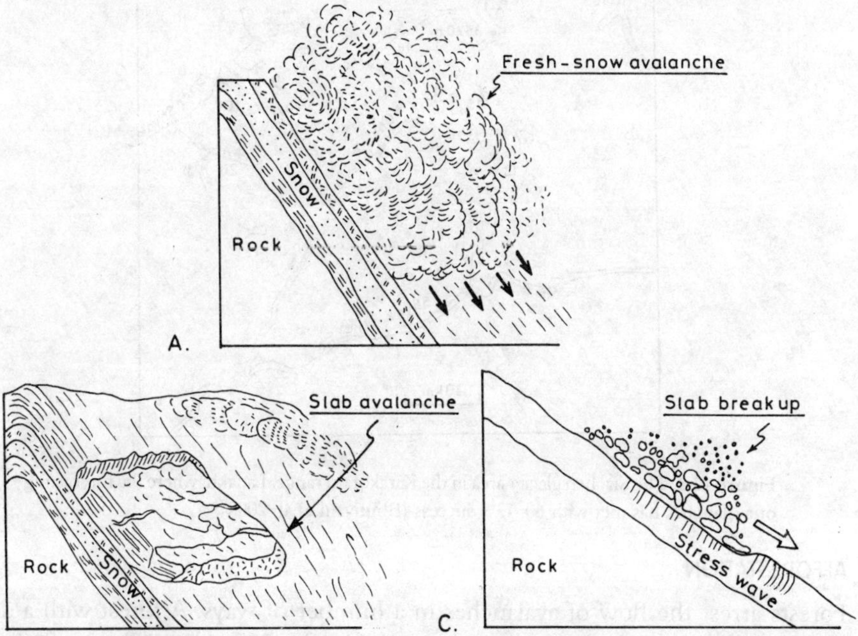


**Figure 12.2.** (A) Fresh snow avalanche, also called direct action avalanche, induced by strong storms, descends as a billowy-powdery mass with speeds exceeding 30 km/hour.
(B) Weakening of ice-crystal bonds in lower layers of the snow mass, and development of fractures at the head, cause slab avalanches. The slab slides down and breaks into blocks and fragments.

## Hazard Mitigation Measures

### FORECASTING

Knowing the place and time of the occurrence of an avalanche helps to avert disastrous consequences along highways. Attempts have been made by the Snow and Avalanche Studies Establishment of the Defence Ministry to predict the time of mass-movement of snow along the Leh–Kargil highway, and in the Siachen glacier in the Karakoram range (Figure 12.3). The algorithm developed for this purpose takes into consideration physical properties such as snow permeability and porosity, and generation of meltwater which is calculated by an energy-exchange process (Bhutiyani et al, 2002). The degree of success in forecasting is 60–75%.

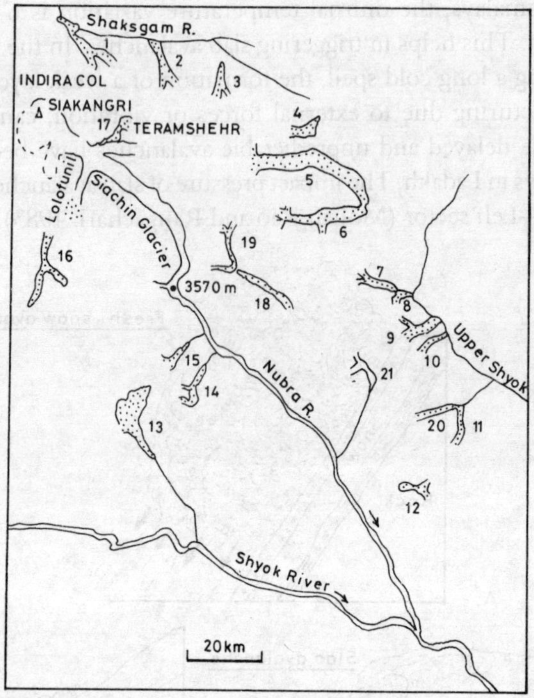


**Figure 12.3.** The Siachen glacier area in the Karakoram range, Ladakh, where the forecasting of avalanches has met with 60–75% success (Bhutiyani et al, 2002).

## AFFORESTATION

Forests arrest the flow of avalanches in a number of ways. A forest with a dense growth of tall trees in the starting zone inhibits the formation of avalanches. Tree trunks support the snow cover and provide anchors to potential slab avalanches. They eliminate snow drift, retaining the snow on tree canopies and releasing it gradually. The forest canopy moderates variability in the net energy-exchange with the snow surface, tending to produce uniform snow temperature distribution and stable snow cover (Sharma, 2004).

## PRE-EMPTIVE ARTIFICIAL TRIGGERING

In December 1987, artillery gunfire during military skirmishes in the Siachen glacier (Figure 12.3) triggered an avalanche that killed 27 soldiers. If one could trigger avalanches at chosen times and under controlled conditions, much of the danger to life and rolling stocks on highways can be averted. An avalanche can be triggered by hurling explosive projectiles or training artillery guns at slopes with thickened piles of snow. Even air guns have been found quite useful. In the Nubra valley in the Siachen glacier, light-weight portable infantry weapons are used to prompt avalanches in order to forestall the possibility of naturally occurring, unpredictable avalanches. On the Jammu–Srinagar highway, at Jawahar Tunnel, avalanches are triggered by firing explosives one or two metres above the snow surface—through a device known as *Avdhav Visphotak Vahan*, developed by the Snow and Avalanche Studies Establishment (Sharma, 2004). This device ensures that the explosive is carried to its desired place of explosion.

## ENGINEERING STRUCTURES

A variety of engineering structures can be constructed in avalanche-prone areas. The structures should be constructed in the formation zone to retain the snow, and in the middle and run-out zone to reduce the intensity and destructiveness of avalanches, by retarding the motion of the avalanche. These structures serve to decrease the speed of avalanches or to divert their courses (Figure 12.4). Structures have also been developed that alter the pattern of snow accumulation in the formation zone, due to wind-drift snow. Below the snow line, **snow fences** (Figure 12.4) built with timber, aluminum or steel tubes prevent any concentration of accumulating snow. Lower down the slope they hold back the snow mass from advancing towards settlements. The **snow baffle** helps to change the course of the avalanche and minimize the thrust (impact) of the main flow. The **avalanche wedge** (Figure 12.4) deflects and breaks the advancing mass of snow, and also diminishes its speed. It is one of the most effective devices to protect transmission-line poles, towers, smaller vital installations and buildings.

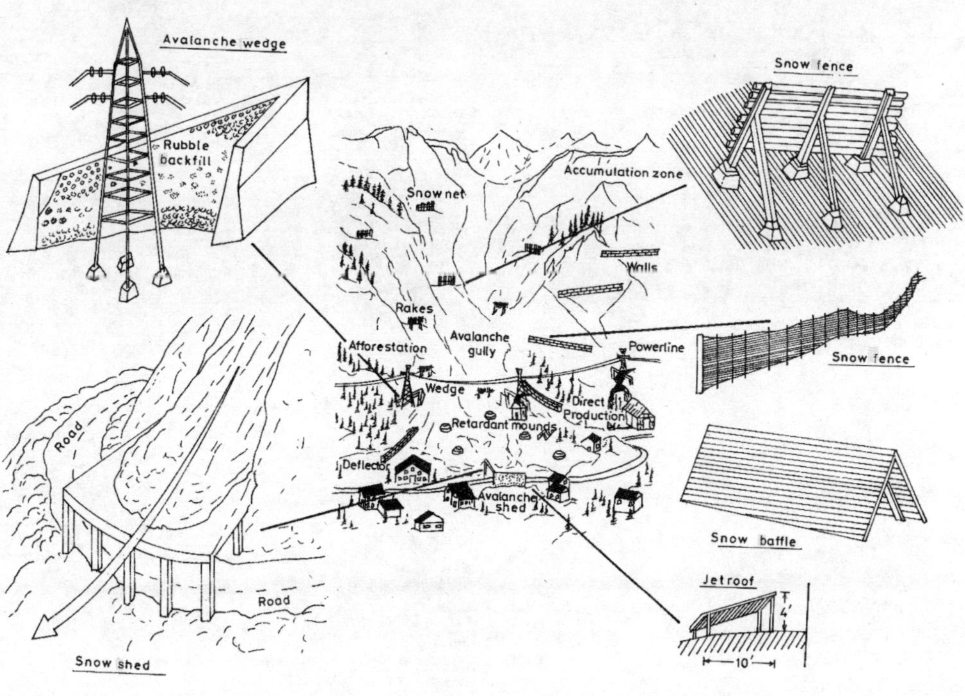


**Figure 12.4.** Various engineering structures that impede the speed of the main flow, or deflect the course of the moving snow mass.

**Diversion dams** divert avalanche-snow away from an object. Retarding structures like catch dams and mounds present a large surface area to flowing avalanches. They help in shortening the run-out distance of an avalanche. The Snow and Avalanche Study Establishment has successfully developed such structures and demonstrated their technology to potential users (Sharma, 2004). The **snow bridge** and snow ramp allow free passage to avalanches over roads and installations, without any damage to the structures. The **jet roof** causes the wind to blow away the cornices and overhangs of accumulated ice. Modified forms of these structures have been used successfully in the Himalaya. It is necessary to emphasize that it is a combination of these engineering structures that could minimize the hazard.

Avalanche hazards have begun to assume dimensions comparable to other natural hazards in India. Thus there is a growing need to generate greater awareness among backpackers, tourists who indulge in adventure sports and local inhabitants. The need to improve the accuracy of avalanche forecasting by developing avalanche forecasting models for different areas is also significant (Sharma, 2004). Although active methods like avalanche control structures and afforestation are expensive, their use is unavoidable where higher levels of protection are needed. Policies at the local level, to create well-equipped and trained Avalanche Safety and Rescue Organizations, are also essential.

# 13

# Enduring Climate Warming and Droughts

## Near-surface Temperature Rise

From various observations, it is abundantly clear that since AD 1900, the earth's near-surface temperature has risen by 0.6°C (Tett et al, 1999), more precisely 0.45±0.15°C (Sahai, 1999). Over the Indian landmass, the mean annual near-surface air temperature (Figure 13.1) has been increasing at the rate of 0.4°C per century (Srivastava and Goel, 1998). The Himalayan province has not escaped this warming phenomenon. There is now reduced snowfall in the Himalaya—high-altitude lakes do not freeze during winter, indigenous plants are giving way to those that grow in warmer climates, and birds seek cooler regions elsewhere.

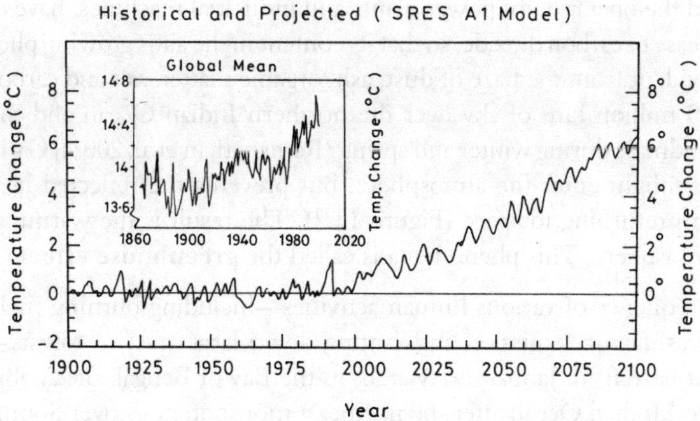


**Figure 13.1.** Annual mean area-averaged temperature over the Indian landmass has been rising, as a simulated mode shows (Lal et al, 2001).

*Inset* shows global climate warming trend in the period 1860–2000 (Brown, 2001).

Recent work done by scientists of the National Climate Data Centre (in Ashville, U.S.A.) shows that the rate of **global climate warming** since 1976 has increased by 2–3°C per century. The rate of global climate warming was 0.6±0.2°C in the last century, and is expected to be 1.1°C in the next quarter of this century. By AD 2100, the surface temperature should increase by 3.5°C (Vaughan, 2001). There is thus an acceleration in the rate of global warming. A modelling study (Lal et al, 2001) predicted that, over the Indian subcontinent, the annual mean area-averaged surface warming would range between 3.5–5.5°C by AD 2080—more perceptibly during the winter season. Warming is perceptible and spatially more consistent in the lower and middle latitudes, and particularly in the tropical oceans. The temperature change is more pronounced during winter and spring, not so much in summer and autumn (Sahai, 1999).

## Causes of Climate Warming

One thing is very clear: the temperature rise in the past fifty years cannot be attributed entirely to the normal internal climate change that has been going on throughout Quaternary time. Various observations and studies demonstrate that this change is the result of a concentration of **greenhouse gases** and **sulphate aerosols** in the atmosphere. These gases and aerosols are the products of human activities—they are anthropogenic. Global warming in the period 1946–96 is undoubtedly human-induced. In some areas, the cooling effects of aerosols are large enough to more than offset the warming caused by greenhouse gases. However, greenhouse gases are longer-living than aerosols, hence the former have more long-term effects (Watson et al, 1998).

Among the greenhouse gases, carbon dioxide ($CO_2$) occupies a crucial position and plays the predominant role. Its content in the air has been rising at a rate considerably more than 4% per decade (Brown, 2001). The burning of mineral-, fluid- and wood-fuels, and the operation of power plants and industrial machines, have contributed to the release of carbon dioxide, so that its content in the air is growing phenomenally. This is evident from the haze of dust, ash, organic matter, salt and carbon that now covers 10 million km$^2$ of sky over the northern Indian Ocean and the southern Indian landmass during winter and spring (Ramanathan et al, 2001). Carbon dioxide lets the sunlight enter the atmosphere, but prevents the reflected heat (infrared rays) from returning to space (Figure 13.2). The result is the warming up of the lower atmosphere. This phenomena is called the **greenhouse effect**.

As a consequence of various human activities—including burning fuel-wood and dung-cakes, factory emission and cutting down large tract of forests—there is a persistent haze from January to March, in the Bay of Bengal, the Arabian Sea and the tropical Indian Ocean after the northeast monsoon rains over South Asia. This has been named the Asian Brown Cloud. Although it is a matter of concern, the Asian Brown Cloud is unlikely to have a substantial impact on crop yields, as projected in a report by the United Nations Environment Programme (Srinivasan and Gadgil, 2002).

Large-scale deforestation has contributed greatly to the greenhouse effect, because of the increase in carbon dioxide in the air. (Plants utilize $CO_2$ during photosynthesis to produce food and oxygen.) With less than 80% of the Indian landmass bereft of forests, it is natural that the amount of $CO_2$ in the air will accumulate. Moreover, oceanwaters are also falling prey to the warming phenomenon, and there is little dissolution of $CO_2$ in the ocean. Oceanwater, at a certain temperature, plays the role of a sink in absorbing $CO_2$ in the air.

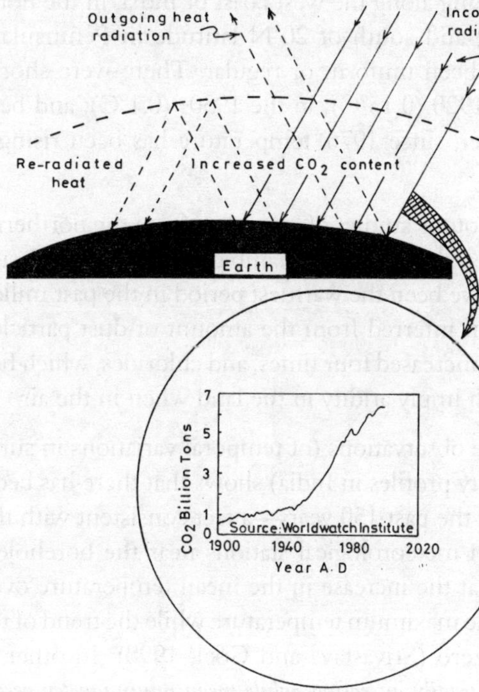


**Figure 13.2.** Greenhouse effect caused by carbon dioxide ($CO_2$). The gas allows the entry of sunrays but prevents the return of the reflected heat rays to space. *Inset* shows progressive increase in $CO_2$ in the air in the period 1900–2000. (*Inset* based on Brown, 2001).

With $CO_2$ are produced gases like nitrous oxide, ammonia and methane, chlorofluorocarbon, hydro-chlorofluorocarbon, carbon tetrachloride, aerosols from petroleum-fired engines (including motor vehicles and diesel locomotives, refineries, pesticides and fertilizer plants), refrigerator-used gases, etc. Nitrous oxides and chlorofluorocarbons have been nibbling away at the ozone layer. The **ozone layer** lies at a height of 16–60 km in the atmosphere and absorbs dangerous ultraviolet rays, preventing them from entering the earth's lower atmosphere. The holes made in the ozone by these gases have allowed ultraviolet rays to reach the earth's surface, which has contributed greatly to global warming. Big, expanding ozone holes over Antarctica and the Arctic—and also over some parts of the Himalaya—imply that the earth is vulnerable to this increasing exposure to ultraviolet rays. In AD 2001, India produced about 18,000 million tonnes of chlorofluorocarbon for codants in refrigerators and air-conditioning systems (N. Singh, 2002). Other ozone depleting substances, like hydro-chlorofluorocarbon and carbon tetrachloride, find extensive use in deodorants, sprays, perfumes, solvents, fire extinguishers and fumigants. In order to stop this depletion of the ozone layer, the manufacture of these products is scheduled to phase out by AD 2010–40.

## Trend of Temperature Rise over the Indian Subcontinent

Despite the remarkable stability of the summer monsoons in the last 120 years (Thapliyal and Kulshrestha, 1993; Srivastava and Goel, 1998; Sahai, 1999), there has been significant warming along the west coast of India, in the northern part of the Indian subcontinent, and south of 20°N latitude in Peninsular India. The temperature rise has not been uniform or regular. There were short intervals of cooling—between 1900–1920 (0.15°C); in the 1930s (0.1°C); and between 1955–75 (Sahai, 1999). However, since 1976, temperature has been rising consistently and rather rapidly.

Oxygen- and hydrogen-isotope studies of ice covers from the northern flank of the Himalaya, bordering Tibet, demonstrate unambiguously that the last fifty years—particularly the 1990s—have been the warmest period in the past millennium (Tett et al, 1999). This has been inferred from the amount of dust particles trapped in Himalayan ice, which has increased four times, and chlorides, which have increased two times—both of which imply aridity in the land when in the air.

Interpretation of borehole observations (of temporal variations in surface ground-temperature, across seventy profiles in India) shows that there has been a warming of about 0.9°±0.1°C over the past 150 years—a fact consistent with the surface air temperature records from meteorological stations near the boreholes (Roy et al, 2002). It was observed that the increase in the mean temperature over the Indian landmass is seen only in the maximum temperature while the trend of the minimum temperature was nearly zero (Srivastava and Goel, 1998). In other words, *mean maximum temperatures are steadily increasing, while mean minimum temperatures have not varied for some time*. Furthermore, the anticipated surface warming is more pronounced during winters. This trend will continue in the future. During the winter season, towards the end of the twenty-first century, the northern and central parts of the Indian subcontinent are likely to experience mean surface warming upto 3°C, while in much of the southern Indian landmass the temperature rise would be less than 2°C (Lal et al, 2001). In the monsoon period, the warming would be approximately 2°C in northern and eastern India, and less than 1.5°C in southern Peninsular India (Lal et al, 2001).

The consequences of global warming are alarming—weed species will slowly displace tree species and the resultant increase in evapotranspiration will have a negative impact on the productivity of crop plants. Wetlands are likely to shrink in size and many areas will face desiccation. *Sorghum* crops visibly showed adverse effects in rainfed fields. Wheat yields would be less than those in the current climate, even with the beneficial effects of $CO_2$ (Rao et al, 1995).

### IMPLICATION FOR RAINFALL

The impact of surface warming will be pronounced to quite perceptible in rainfall patterns over the Indian subcontinent. During the winter season, there will be 5–25% less rainfall (Figure 13.3), while in the summer monsoon, the rainfall will be

10–15% more than the normal (Lal et al, 2001). This will not only affect groundwater resources and farming, but also make the regions of eastern and northeastern India vulnerable to greater floods. As a consequence of the decline in winter rainfall, there will be drought during summer, until the arrival of the monsoon rain—which is likely to become variable, especially irregular over central India. There will be more spells of intense rainfall, and more frequent instances of extreme rainfall (Lal et al, 2001).

The change in climatic conditions will have a profound impact on agriculture. In the global context, a rise of 1.4–5.8°C by AD 2050 would result in upto 30% decrease in farm output (Brown, 2001).

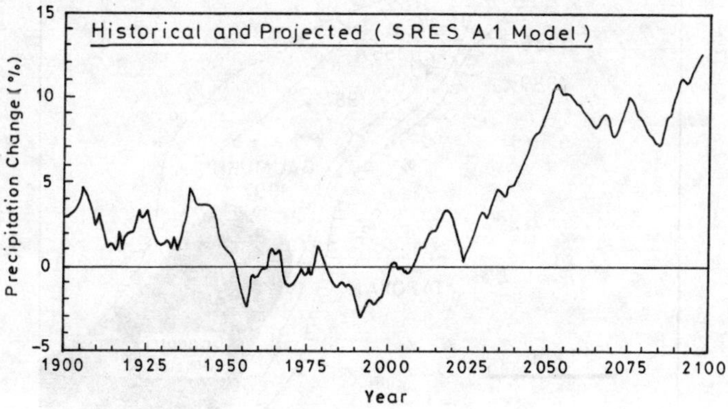


**Figure 13.3.** Anticipated trend in rainfall change (in % of ten-year running mean) over the Indian subcontinent due to the rise of near-surface temperatures (Based on Lal et al, 2001).

## Melting of Ice Caps and Glaciers

Caps of ice and snow—covering high mountains in the continents, and Antarctica and the Arctic in the polar regions—are fast melting. This is evident from satellite images and many ground-based studies. The melting increased considerably in the 1990s, evident from rising water levels in at least 44 glacial lakes in Nepal and Bhutan. This has, incidentally, increased the threat of floods from glacial-lake bursts. As already stated, the 1990s were the warmest decade in the millenium. The Antarctic ice pack melted considerably along the edges, with some shelves disintegrating disastrously. The western ice sheet shrank at the rate of 122 m/yr. The thickness of the Arctic ice pack, likewise, fell from the average 3.1 m to 1.8 m in less than 30 years, and its spread shrank by 6% in the period 1978–96. The loss of Arctic ice was 34,300 km$^2$ in less than 20 years. Analysis of satellite data over the period 1978–2002 shows that Arctic ice cover will disappear within this century (*Down to Earth*, 31 December 2002).

Mountain glaciers are retreating all over the world. At the present rate of **glacial recession**, nearly 25% of the world's glaciers will disappear by AD 2050 and the Himalayan ice cover will shrink to one-fifth of its present extent by AD 2035! The Gangotri glacier—the source of the Ganga in Garhwal—has been consistently retreating at a rate that can be called rapid, except for a short period between 1890–1910 (Figure 13.4). It retreated at the rate of 46.4 m/yr in the period 1962–90, 37.4 m/yr during 1990–94, and 25 m/yr in between 1994–98 (Tangri, 2000). The Pindari glacier (in Uttaranchal) retreated 2,840 m in 121 years (1845–1966); the Milam glacier (in Uttaranchal) by 1,350 m in 108 years (1849–1957); and the Zemu glacier (in Sikkim) by 44 m in 56 years (1909–1965) (Vohra, 1978; 2000). As a matter of fact, practically all Himalayan glaciers are retreating (Plate 19). With the retreat of these glaciers, the vast ice cover over the Himalaya is shrinking.

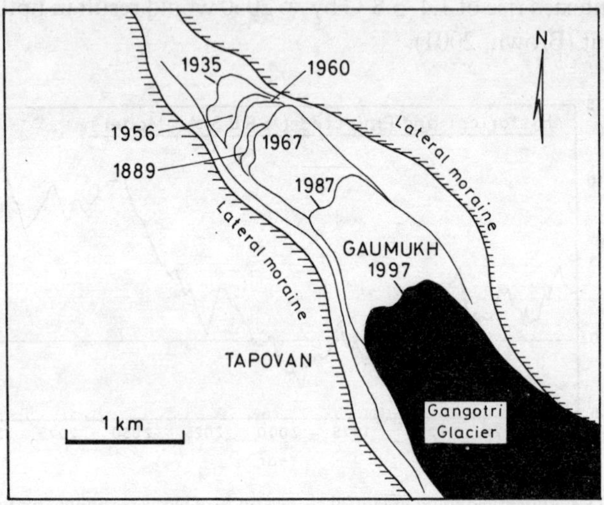


**Figure 13.4.** The Gangotri glacier, from which the Ganga emanates in Garhwal, has been consistently retreating from at least 1910 onwards (After Mukherjee and Sangewar, 2001).

## FALLOUT OF ICE-MELTING ON THE OCEAN REGIME

Increased melting of ice and snow implies greater availability of water in the short term, but a decrease in the long term. This scenario leads to a reduction in the average flow, but an increase in the peak flow—which has an appreciable impact on the generation of hydroelectricity, and on agriculture dependent on water from the snow-fed Himalayan rivers.

The melting of ice on a stupendous scale means there will be more freshwater in the oceans. This will cause reduction in the salinity of oceanwater which will in turn impact aquatic life conditioned to the existing level of salinity. As more and more meltwater is added to the oceanwater, aquatic life will be forced to adapt to the changed level of salinity or else perish. The cooling of oceanwater will also have an effect on life under the sea.

Since meltwater could cause lowering of temperatures in the Arctic and Antarctic oceans, the temperature gradient that exists between tropical and high-latitude oceans will alter and bring about a change in the circulation of ocean currents. This development is bound to modify the global climate considerably.

### RISE IN SEA LEVEL

The melting of ice caps and glacial snow has raised sea levels perceptibly. According to one estimate, in the last one hundred years, the sea level has risen by 10–25 cm, while another estimates it at 20–30 cm. The global mean sea level rise is variable. The Mumbai tide gauge data, for the period 1878–1997, indicates that the sea level rose at the rate of 0.02 mm/yr (Watson et al, 1998). According to Survey of India measurements in the last 25 years, the rate of sea level rise along the Indian coast is between 0.5–11 mm/yr. The rate of sea level rise today is thus ten times higher than what it was in the past 3,000 years. A modelling study predicts that by AD 2050, the sea level would be 21 cm higher than what it is today, and by the end of the century, it will stand 100 cm higher than the present level.

Airborne laser altimetry of 67 glaciers in Alaska has showed that, between 1955–95, the average rate of decrease in the glaciers' thickness was 0.52 m/yr, while volume reduction was $51 \pm 15$ km$^3$/yr. In contrast, between the mid-1990s and 2000–01, the thickness decreased at an average rate of 1.8 mm/yr, while the volume shrank at the rate of $96 \pm 35$km$^3$/yr. This fact implies that the rate of sea level rose by $0.14 \pm 0.04$ mm/yr between 1955–95, and $0.27 \pm 0.10$ mm/yr between the mid-1990s and 2000-01 (Arendt et al, 2002).

The consequence of such a rise in sea level will be far-reaching and grave. One metre rise in the water level means displacement of more than 7.1 million people from the submerged tracts—in nearly 6000 km$^2$ or 576,400 ha of coastal land in the Indian subcontinent, along with 4,200 km of coastal roads. In West Bengal and Orissa, the submergence of 170,000 ha of land will uproot 700,000 people. Goa stands to lose 4.32% of its coastal land. In Bangladesh, nearly 15 million people will be affected if seawater inundates 30,000 km$^2$ or 17% of its land (Watson et al, 1998).

Following the rise in sea level, low-altitude islands will be completely submerged Tuvalu, a small island in the Pacific Ocean, is likely to disappear within a few years, and the Maldive islands in the Indian Ocean will be the next victim of the deluge.

The sea level rise may cause either landward migration of mangroves and tidal wetlands, or their submergence and eventual effacement. Not only would there be acceleration of erosion in the coastal belt due to the effacement of mangroves and sand dunes, but also increased hazards from tropical cyclones. The storm surges that rise 6–8 m high would spread over much larger parts of the low-lying areas. Thus large tracts of cultivated land would be seriously threatened by the saltwater that will come in the wake of storm surges.

## Climate Warming through Geological Ages

Climate has been changing through the geological ages, although the change has been variable in intensity and periodicity. The last 10,000 years of the Holocene epoch can be used to illustrate this development. The cold dry climate of the Ice Age persisted until about $10{,}500 \pm 500$ yr BP. This was followed by a period of warmth and wetness, when rainfall increased. In the arid–moist transition period (12,000–8,000 yr BP) the moisture in the air, among other things, checked the spread of sand dunes in the western and northwestern parts of the subcontinent. Study of pollens trapped in sediments of lakes, lagoons and swamps in various parts of the country, as well as the ice cover of the Himalaya, and oxygen- and hydrogen-isotope studies of groundwater in Tamilnadu (Sukhija et al, 1998) point to a period of heavy rainfall between 10,500–3500 yr BP, particularly between 6000–4000 yr BP (Valdiya, 2002). This was followed by a rise in temperature and the prevalence of aridity in the period 3500–2000 yr BP. Since, then the temperature has been rising. However, for a short period of 400 years between AD 1450–1850, temperatures fell and glaciers advanced in the mountains (known as the Little Ice Age).

The superimposition of warming on the natural internal rise of temperature, thanks to human activities like deforestation and industry, has aggravated the situation, particularly in the vulnerable areas. The most vulnerable ecosystem is the soil, which is losing interstitial moisture (Figure 13.5 *inset*).

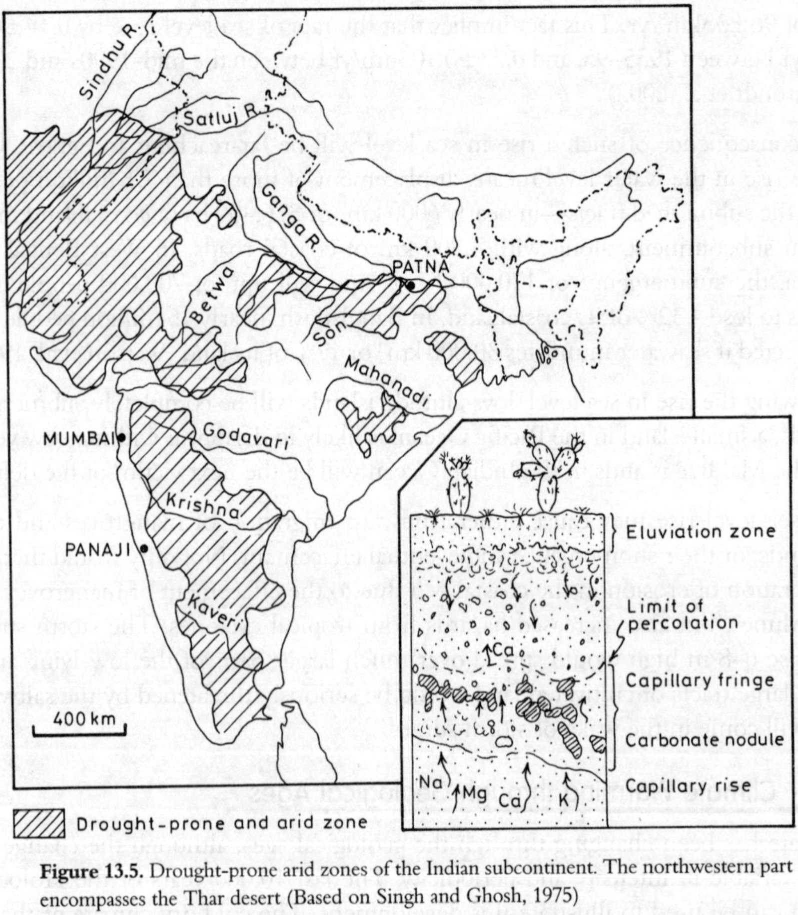


**Figure 13.5.** Drought-prone arid zones of the Indian subcontinent. The northwestern part encompasses the Thar desert (Based on Singh and Ghosh, 1975)

Inset shows formation of aridisol in the arid zone, where there is fast or total destruction of organic matter in the soil. Capillary action brings up salts and carbonates from the deeper level. Only xerophytes survive in such a situation.

### MOISTURE-LESS SOIL MICROCOSM

As already stated, the soil is a microcosm of sorts, and the dwindling or disappearance of water vapour and the vapourization of interstitial water leads to the death of the organisms that live within it. This situation renders the soil incapable of supporting plant life. The soil then becomes a dead ecosystem, in a situation that is described as **soil drought**. The onset of soil drought is preceded by extreme deficiency, or total absence, of rainfall for prolonged periods. A reduction in rainfall, below a specific amount for a specific period, is described as **meteorological drought**. Reduction of water resources—as seen in stream flow, lake level and groundwater falling below a specific level—is called **hydrological drought** (Gupta et al, 2001).

The development of drought over a widespread region is a slow process. It may persist for months or for years. There is a drought of moderate intensity once every three years, in some part of India. Severe droughts occur, on average, every 32 years. The worst drought in the period between 1789–1993 killed 600,000 people. Very severe droughts (that took a heavy human toll) occurred in 1877, 1899, 1918, 1941, 1951 and 1972. More than 16% of the geographic area of India—or 68% of the total sown area—is prone to droughts of varying degrees (Figure 13.5), affecting more than 50 million people annually. Apart from erratic and inadequate rainfall, the late onset and/or early withdrawal of the summer monsoon is also responsible for the development of drought conditions. The El Nino event in the oceans is known to have a perceptible effect on rainfall in the northern and central parts of the Indian subcontinent, usually from a bad monsoon (Sikka, 2004). This is evident from Table 13.1.

**Table 13.1.** Low rainfall related to El Nino effect (After Gupta, 2002).

| El Nino year followed by Drought | Standard Rainfall Anomaly |
|---|---|
| 1877 | – 3.0 |
| 1899 | – 2.7 |
| 1905 | – 1.6 |
| 1911 | – 1.4 |
| 1918 | – 2.4 |
| 1928 | – 1.0 |
| 1941 | – 1.5 |
| 1951 | – 1.4 |
| 1965 | – 1.7 |
| 1972 | – 1.3 |
| 1982 | – 1.4 |
| 1987 | – 1.3 |

## Meeting the Challenge of Drought

The strategy for combating the menace of drought entails: (i) changing grazing practices by restricting grazing to fenced areas surrounded by fodder plantations, (ii) stabilizing soil using biological and technological methods, (iii) harvesting rainwater for varied needs, and (iv) treating soil sickness caused by alkalinity–salinity with appropriate use of technology, including construction of drainage networks (as described in Chapter 4).

Water harvesting, as elaborated in Chapter 5, is doubtless the most effective and easy method of combating the drought problem. Rainwater runoff should be gathered through converging channels and diverted to settling basins. The collected water must then be led through pipes or underground tunnels to cisterns or storage holes dug below the ground. Adequate measures should be taken to prevent loss from seepage and evaporation. Floodwater, likewise, can be diverted and stored in depressions on land. Improving water supply and water conservation measures entails efficient runoff management through drainage systems. This is to be backed by measures for soil conservation, tree plantation, livestock management, and appropriate distribution of water.

# 14

# Applying Careful Technology for Development

## Tenets of Environmental Geotechnics

All activities related to the development of facilities—the provision of water for various purposes, the generation of energy, the building of different modes of transportation and communication, etc—invariably bring about changes in ecological systems. In most cases, the changes are drastic and bring about an imbalance in the environmental order. Even 'minor tinkering' can affect the whole system. These imbalances may be aggravated to eventually become the principal cause of hazards. In pursuing development activities, one cannot help recalling the ringing words of Leopold Aldo (1941),

> Man, too wise to tolerate hasty tinkering with our political constitution, accepts without qualm the most radical amendments to his biotic constitution.

It is therefore imperative that civil engineering practices are governed by a commitment to abstain from causing harm to ecological systems. Not only that, every engineering activity must be invested with a social meaning. This means encouraging dialogues between what Braun (1991) called 'the risk creators' and 'the risk bearers'. In other words, there has to be effective communication between planners, engineers, environmentalists, geologists and social scientists, and the local people during the planning and implementation of development programmes.

The purpose of engineering geology and geotechnics is to ensure that no part of an ecological system is hurt beyond a tolerable limit and that harmful effects are kept to a minimum—underlining the fact that the methodology of geotechnics must not become a careless technology. This calls for a comprehensive appraisal of the impact of engineering work on the environment and social structure, *before* embarking on a project. In other words, an **Environmental Impact Assessment** must be carried out parallel with geotechnical investigations for every development programme.

### GEOTECHNICAL INVESTIGATION

The objective of a geotechnical investigation is to understand the natural conditions that affect the safety, the soundness and the economics of an engineering project to be undertaken. In rocky terrain, the design and size of a project depends on the lithological and structural setting of the area. The integrity and safety of a structure depends on the extent and geometry of structural discontinuities such as cleavages, joints, fissures and faults, and on the nature, thickness and disposition of the rocks and the overlying soil. In this context, the bearing capacity and shear strength of the foundation are important parameters that govern the soundness, safety and economics of a civil engineering structure to be constructed.

**Table 14.1.** Methodology of site investigation for water resources development (Based on Chaplow, 1981).

| Phase of Investigation | Geotechnical Investigation |
|---|---|
| (1) Preliminary Study | (a) Collection and review of existing data on the region's geology using satellite imagery, topographic maps, aerial photographs, reports, etc.<br>(b) Site reconnaissance<br>(c) Commencement of geological mapping<br>(d) Limited subsurface investigations |
| (2) Feasibility Study | (a) Detailed geological mapping<br>(b) Subsurface site investigation<br>— geophysical surveys<br>— borehole observations<br>— exploratory excavations<br>— *in situ* testing<br>— laboratory testing<br>— instrumentation<br>(c) Commencement of special studies (e.g. monitoring seismicity) |
| (3) Design Study | (a) Completion of geological mapping<br>(b) Selective detailed geological mapping<br>(c) Conducting subsurface site investigation<br>(d) Continuation or conclusion of special studies (e.g. monitoring seismicity)<br>(e) Laboratory investigations for collecting data concerning designs of dams, etc. |
| (4) Construction Investigation | (a) Geological logging of all excavations<br>(b) Assessment of geological predictions of earlier stages as the excavation progresses, and review of the recommendation based on actual field conditions<br>(c) Other investigations as found necessary during construction |
| (5) Operation Monitoring | (a) Recording and interpreting the results of project instrumentation<br>(b) Visually monitoring behaviour of project area, specially slopes.<br>(c) Performance evaluation of the project by interpreting the results<br>(d) Environmental impact studies |

**Geodynamic conditions** of a region—such as seismicity, neotectonism and the flow-pattern of groundwater—also have a profound bearing on designing structures and locating their sites. It is also important to know if the chosen area is prone to any kind of natural hazard. Regardless of delay in the implementation or cost of construction, these two aspects must be assessed and addressed properly. Thus, advance geological information is an indispensable pre-requisite for any cost-benefit analysis. A model of this methodology is given in Table 14.1.

An important item in any geotechnical investigation is the identification and precise delineation of the zone, area or spot of potential hazards (earthquakes, landslides, ground subsidence, soil creep, floods, etc). This knowledge helps to make the necessary changes in the location, design and size of the structure planned. A few examples will bring home the point. At the Pong dam across the Beas river in Himachal, the intake structures for the tunnel had to be shifted to the left abutment-hillslope to cope with the risk of slope failure. In the Giri river in southeastern Himachal, a powerhouse had to be constructed underground taking into account the endemic landslides triggered by a shear zone related to the Main Boundary Thrust. The cavernous nature of Kalhel limestone in the Siul valley (a Ravi tributary), forced the abandonment of the project altogether. In the Pench valley in Maharashtra, the Kamthikheri project site was strengthened by resorting to intensive grouting of the cavernous marble foundation.

In mountain domains, road alignments should be laid and rock-cutting for the construction of buildings undertaken only after ascertaining the stability of the hill slope (Chapter 9), and making sure that groundwater circulation will not be intercepted or adversely affected.

## Cost-benefit Analysis

Needless to say, every possible action on the environment and its resources has a cost. It is therefore necessary to compute the monetary values of the loss of resources—both living and non-living—and also the cost of remedial measures that will have to be taken to restore the environment to as much of its original state as possible. The resources that have a price include agricultural fields, forests, pastures, wildlife, aquatic fauna and mineral deposits. Some remedial measures pertain to: (i) stabilizing slopes due to excavations for roads, canals, buildings, mining, (ii) recharging groundwater following its interception/disruption due to cutting of slopes, (iii) tree planting following loss of forests due to construction activities, and so on. The damage to or the destruction of agricultural fields, forests, water bodies, etc, brought on by landslides caused by mining and road-cutting, needs to be compensated with ample monetary packages. The submergence of agricultural land under reservoir water behind dams, or waterlogging of productive land due to leakage from canals or over-irrigation, which leads to diminished yield, are to be taken into consideration in calculating the costs of a project.

There are also intangible fallouts, such as the mental distress or trauma which affects people who are uprooted from preferred locations or environments. There is no way to cost this factor. In making an appraisal of the worthiness of a project, it is essential to seek the viewpoint of the people who would be affected by the development project.

## Environmental Impact Assessment

An Environmental Impact Assessment (EIA) (Table 14.2) is carried out to ensure that a project under consideration is environmentally sound and would not stand in the way of sustainable development. A pre-requisite for an exercise in EIA is the gathering of information on: (i) the effects of the project on the environment, (ii) the risks that are likely to develop, and (iii) development options and their consequences/implication.

1. EIA uses scientific data to articulate: (a) the socio-economic significance of the project, and (b) a commitment from planners/developers stating that the expected benefits will be shared between the communities for whom the project has been conceived and the people who would be adversely affected by it. It may be stressed that monetary compensation for the loss of land, property, home, etc cannot be construed as sharing of benefits—it is no more than a futile attempt to buy peace, that too temporarily.
2. The EIA document should expose the full gamut of environmental changes likely to occur as a consequence of engineering activities related to the implementation of the project.
3. The EIA report must state in clear terms, the sociological, cultural and psychological problems that are likely to arise due to any displacement of people, and/or the disruption in life of the community.
4. Any EIA should evaluate the financial burden (the cost) of: (a) restoring the natural environment to its original state as far as possible, and (b) resettlement of the displaced people, which includes providing them with all essential facilities.
5. It is desirable to discuss alternatives to the proposed project, pointing out the different ways to realize the objectives in a socially sound, technically feasible and ecologically desirable way. The motto of the project should be Development Without Destruction.

**Table 14.2.** Methodology of Environmental Impact Assessment of development projects.

A. *Proposed Project*
  1. Objectives of the Project
  2. Location and Land Requirement
  3. Existing and/or Proposed Facilities (Cultural and Technical)
  4. Construction Procedure and Schedule
  5. Procedures of Operation and Maintenance, and Safety Measures Proposed

6. Planning Constraints
7. Possible Alternatives to the Proposed Project

B. *Existing State of the Natural Environment*
1. Land Features (Topography) and Land Use
2. Geological Setting and Condition of Foundation
3. Geological Hazards
4. Meteorological Condition and Hydrologic Regime
    (a) Climate and Rainfall
    (b) Surface Hydrology
    (c) Underground Hydrology
    (d) Water-quality and its Monitoring
5. Biotic Resources
    (a) Vegetation
    (b) Wildlife
    (c) Fish and Aquatic Communities
    (d) Genetic Resources
    (e) Endangered Species
6. Non-renewable Resources
7. Cultural Resources (Historical and Archaeological Heritage Sites)
8. Criteria for Measuring Impact

C. *Socio-economic Conditions*
1. Population and Occupational Status, and Demographic Impact
2. Employment Pattern: Existing and Expected
3. Economic Condition and Impact
4. Social Values and Amenities, and Health Conditions
5. Problems of Resettlement
6. Influx of Labour and Consequent Socio-economic Problems

D. *Environmental Impact of the Proposed Action*
1. On Land Features and Uses
2. On Geological and Social Conditions
3. On Ecosystems and Species (Terrestrial, Aquatic and Endangered)
4. On Air and Water
5. On Renewable Natural Resources
6. On Waste Production and Disposal

E. *Measures to Avoid or Mitigate Adverse Environmental Effects*
1. Preventive Measures
2. Remedial Measures

F. *Alternatives of the Proposed Project*
1. Feasibility of Alternative Project
2. Economic Comparison through Cost-benefit Analysis

G. *Recommendations*
1. Significant Environmental Impacts
2. Mitigation Measures
3. Alternative Projects

# 15

# Laying Foundations of Buildings and Utility Structures

## Innate Strength of Bedrock and Soil

People endeavour to construct buildings and utility structures that can withstand the stresses of unequal settlement, ground subsidence, and the forces of floods, cyclones, storms, landslides and earthquakes. The resultant forces that operate in such hazardous phenomena are transmitted to the ground through these very foundations. In the ground underlying the foundation, two types of pressures are generated: (i) **compressive stress** that tends to reduce the volume of the material, as the forces act towards each other, and (ii) **extension stress** that tends to increase the volume of materials, as the forces act away from each other. The latter is known as **shear stress,** which results in deformation, causing sliding or slipping on the plane of weakness, or even plastic flow (Figure 15.1). Since the deformation caused by extension stress is negligible, it is generally not considered in an analysis for designing and constructing buildings and structures.

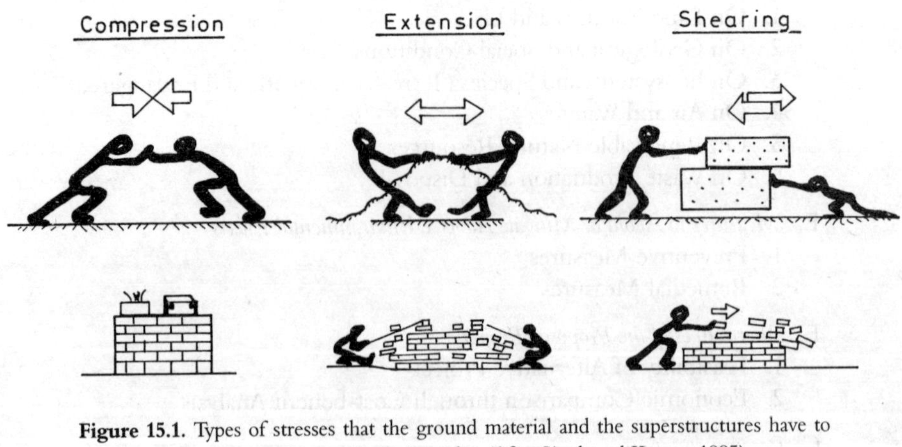


**Figure 15.1.** Types of stresses that the ground material and the superstructures have to withstand when forces of nature come into play. (After Singh and Kumar, 1995).

If the compressive and shear strengths of the ground material beneath the foundation—rocks and/or soils—are strong enough to resist the deformation brought about by compression and shearing stress, buildings and other structures remain stable. Geological investigations throw light on the bearing of bedrocks and/or soils and indicate whether the formation has the ability to support the structure without deformation.

The **compressive strength** of ground material comes from the resistance that its constituents (rock and soil) offer to compression, which are forces acting towards each other. Rocks like conglomerate, coarse sandstone, granite, etc have greater compressive strength than fine-grained rocks like claystone, mudstone, siltstone, etc. The presence of clay and organic matter in material beneath the foundation imparts a degree of instability owing to its greater compressibility.

The following five categories of rocks are recognizable on the basis of compressive strength:

*Very high compressive strength*: finer-grained gabbro, basalt, dolerite, quartzite

*High compressive strength*: medium-grained granite, diorite, basalt, quartzite, compact sandstone

*Moderate compressive strength*: average sandstone, coarse-grained granite, gneiss, amphibolite

*Low compressive strength*: poorly cemented sandstone, shale, phyllite, schist

*Very low compressive strength*: very poorly cemented sandstone, siltstone, mudstone, chalk, volcanigenic tuff

## THE CRITICAL FACTOR: BEARING CAPACITY

The most critical factor in locating, designing and laying the foundation of a structure is the **bearing capacity** of the material that forms the ground beneath the structure. Bearing capacity is the measure of the load or pressure that develops under the foundation—it should be strong enough to prevent damaging displacement or movement in the foundation. Problems arise when there is excessive settlement, leading to deformation of the structure. This deformation manifests itself in the development of cracks and distortion of parts, eventually leading to collapse of the structure. When (or where) the load of a structure is increasingly applied to ground material, the material beneath the structure compresses downwards—forcing the material to move downwards and outwards, resulting in elastic bulging and distortion. Further increase of the load may cause **plastic flow**, which manifests itself in outward displacement and shearing of the ground material (Figure 15.2).

The uplift force that develops as a reaction to loading can be countered by increasing the depth of the foundation beneath the ground surface—the deeper the foundation, the greater the resistance to uplift. Uplift force can also be checked by placing loads on the passive zone (Figure 15.2A). To ensure integrity and stability of a structure, it is essential that: (i) the bearing stress (load) of the structure is kept lower than the bearing capacity of the ground material (Table 15.1), and (ii) settlement within the ground material is made less than the allowable limit of settlement of structures.

**Table 15.1.** Net safe bearing pressure of some rocks (Based on IS Code 12070, 1987).

| Ground Material | Bearing Pressure (tonnes/m²) |
|---|---|
| Granite, diorite, basalt, gneiss, quartzite | 1,000 |
| Schist, phyllite, slate | 400 |
| Limestone and dolomite | 400 |
| Sandstone and shale | 250 |
| Soft and/or broken bedded rocks | 100 |
| Soft shale, claystone | 40 |

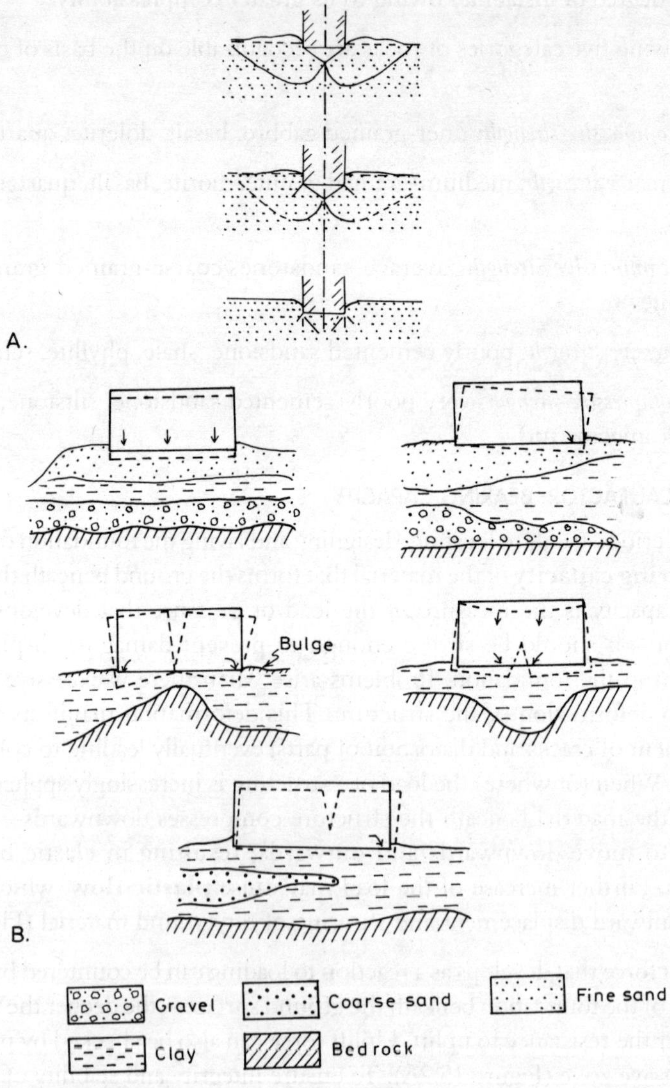


**Figure 15.2.** (A) Failure of the foundation material beneath a structure: (i) total shear failure, (ii) penetration failure.

(B) Layers with varying compressibility, and uneven profile of interfaces, generally suffer from differential settlement (After R. Sanwal, per. com., 2002).

As shown in Table 15.1, igneous, metamorphic, and some sedimentary rocks are considered good foundation material. However, compositional, textural and structural heterogeneity, and structural discontinuity (like fractures, joints, etc) in these rocks may render them vulnerable to differential settlement, sometimes leading to failure. Lacustrine and fluvial clays, water-saturated colluvial deposits, loose aeolian sands and uncompacted landfill deposits are least suitable. While soft clays may suffer large settlement, stiff clays are affected by fissures and cracks (R. Sanwal, per. com., 2002). Although silt is a poor foundation material, loess in its natural and dry state has a low to moderate bearing capacity. A well-designed and compacted sand/gravel fill has the capacity to bear the load of larger structures, but a poorly compacted clay/garbage fill is vulnerable to disastrous settlement.

## FACTORS THAT DIMINISH SHEAR STRENGTH

Shear strength refers to the resistance offered by foundation material to shearing stresses, which tend to move one part from another due to the material's properties of cohesion and the angle of internal friction. Cohesion is the tendency of constituent particles to stick together. Frictional resistance is the innate resistance offered during interaction between individual particles which make up the plane of separation. The presence of water, depending on the porosity and permeability of the material, reduces its frictional resistance and therefore its shear strength. Fine-grained rocks have relatively greater shear strength than coarser-grained ones. Structural features such as open cleavages, foliation, fissures, fractures and joints significantly decrease the shear strength of the rock. This is because the planes of discontinuity allow slipping or sliding, besides admitting water to seep along. Water, as already stated, reduces frictional resistance considerably.

## DIFFERENTIAL SETTLEMENT

Unequal or differential settlement of the ground is the most serious problem in laying foundations for buildings and utility structures. If the extent of settlement is more than 25 mm (more than 12 mm in shallow foundations) there may be serious distortion of superstructures. The settlement of the structure manifests itself in the cracking and deformation of the structure, as already stated. Clays are the most vulnerable to this phenomenon. The settlement may continue for years if it is under persistent shear stress or load. It may be recapitulated that foundation settlement (Figure 15.2) can result from a number of causes, like: (i) load of the structure exceeding the critical limit—the bearing capacity of the foundation material, (ii) groundwater fluctuation, (iii) textural variation (change of grain size) in the sediments or soils that form the foundation material, (iv) structural inhomogeneity or discontinuity, (v) shrinking or swelling of expansive clays within foundation material, (vi) corrosion of the foundation, (vii) subsidence of the ground surface due to factors like underground mining, etc, and (viii) failure of hill slopes.

Floodplains such as the Indo-Gangetic Plain, the Sabarmati basin, the Kaveri basin, and the palaeolakes (like the Karewa in Kashmir and the Kathmandu in Nepal) are characterized by a high content of clayey or muddy material. These areas, therefore, experience persistent problems of foundation settlement. Cavernous limestones are very prone to this pernicious problem. Soft hollow material of fault zones are as incompetent as clays and muds, and therefore prone to foundation problems. Fault zones are unsuitable for laying foundations, particularly for major structures like dams. In the case of active faults, the problem is doubly serious.

## SHRINKING-SWELLING CLAYS

The lava plateaus—Malwa and Deccan—have thick carpets of a unique soil called *regur*, also called **black cotton soil**, characterized by a high proportion of swelling clays. The thickness of *regur* varies greatly (30–15 cm) from area to area. This clay swells in the presence of water—in drier times, the surface of the *regur* is stiff, impenetrably hard and breaks along deep cracks as it shrinks. In the rainy season, the clay expands substantially, giving rise to a very sticky, plastic mass that swells. The swelling is manifest in the formation of mini ridges and bulges on the ground. Since there is a drastic change in moisture content between the summer and rainy seasons, the swelling is also quite appreciable.

Apart from rainwater, water from canals, tanks, pools, sanitary (septic) tanks and toilets causes swelling of clays in adjoining areas. The swelling and shrinking of *regur* soil generates pressure on the foundation of buildings and structures. This results in damage to the structures (Figure 15.3), as witnessed all over the Deccan, including Maharashtra and adjoining parts of Madhya Pradesh, Andhra Pradesh and Karnataka.

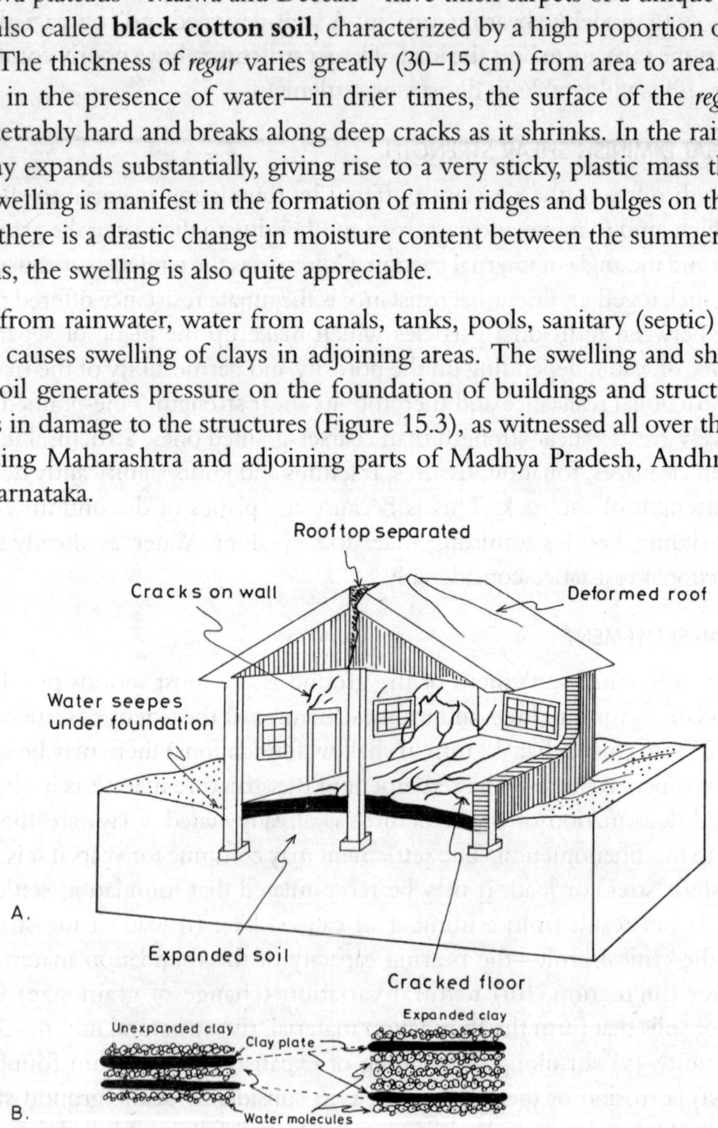


**Figure 15.3.** (A) Swelling of montmorillonite clay within the soil (that constitutes the foundation of the house) has resulted in the cracking of the floor and walls, the parting of the roof and the distortion of door and window frames.

(B) Crystal structure of the swelling clay mineral.

## Appropriate Foundation Types

Realizing the problems that differential settlement and swelling of ground material generate, it would be prudent to design a foundation based on the conditions of the ground. The factors that are to be taken into consideration are the **bearing capacity, allowable settlement** and **compressibility** of the ground material beneath the foundation.

(i) If the ground beneath the proposed foundation is made up of soft, weak, plastic, yielding material—with very low bearing capacity—it is desirable to construct a foundation with a continuous concrete slab that covers the whole area. This is the **mat foundation** (Figure 15.4A). If the situation is extremely delicate, the mat foundation can be developed into a box-like structure (Figure 15.4A). This would act as a **raft** of sorts on the yielding material.

(ii) Where the soil or rock mass of the ground is weak, with low compressive strength, a **spread foundation** (Figure 15.4B) is better suited to cope with the problem. It takes up the weight of the entire structure, and distributes it over a larger area so that the per unit area pressure does not exceed the critical limit of failure.

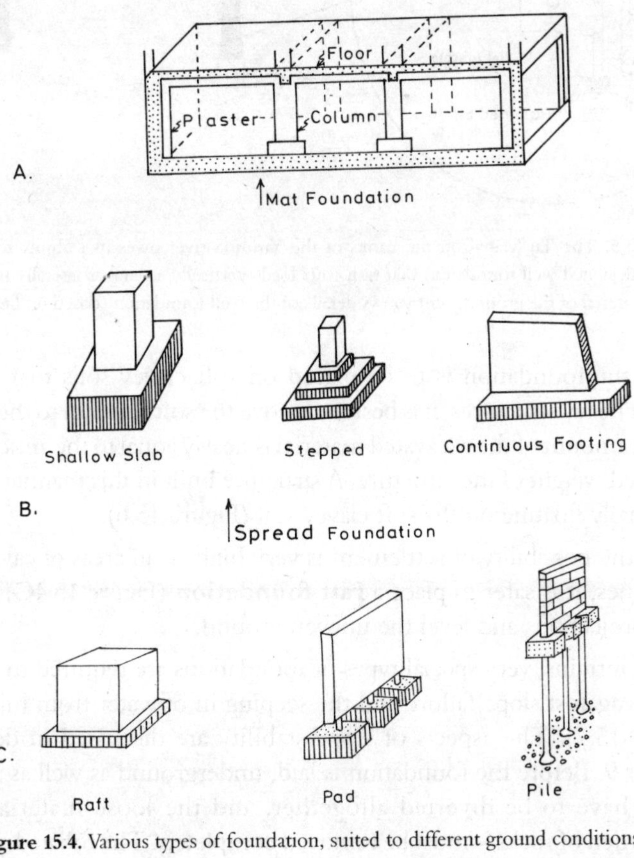


**Figure 15.4.** Various types of foundation, suited to different ground conditions.

(iii) If the ground material comprises a thick aggregate of loose sediment-like gravel, debris or landfill, a **pile foundation** (Figure 15.4C) or a **well foundation** provides maximum stability to the structure. To make this kind of foundation, holes are drilled or deep wells dug and then filled with concrete and reinforced with steel bars. These structures transmit the load vertically downwards into the hardened horizon, and also laterally to the material on the periphery of the holes or wells. The Taj Mahal, on the banks of the River Yamuna, is built on thick unconsolidated alluvial sediment. Resting on a superbly designed well foundation, this architectural marvel has withstood the ravages of time and recurrent floods for more than 300 years (Figure 15.5). Despite the ground material being prone to differential settlement, the Taj Mahal does not show any sign of structural distress.

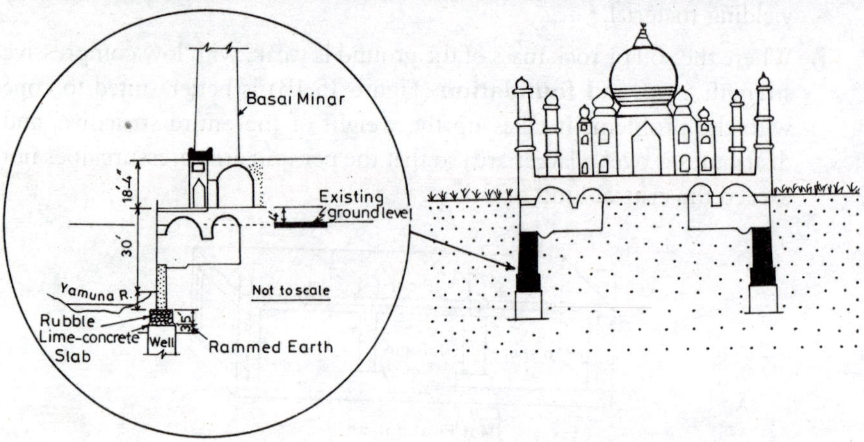


**Figure 15.5.** The Taj Mahal on the banks of the Yamuna river owes its stability to the superbly designed well foundation that transmits loads vertically as well as laterally to the alluvial material of the ground. *Inset* shows details of the well foundation (Based on Legget, 1962).

(iv) Where the foundation is to be placed on soft clayey soils that cannot support heavy structures, it is best to remove the soft material to the extent that the amount of the excavated material is nearly equal to the anticipated/estimated weight of the structure. A structure built in this manner would be virtually floating on the soft clayey soil (Figure 15.6).

(v) Where the possibility of settlement is very high, as in areas of cavernous limestones, it is safer to place a **raft foundation** (Figure 15.4C) on the rocky projections and level the uneven ground.

(vi) In hilly terrains, very special types of foundations are required to ensure stability against slope failure and the seeping in of water from hillslopes (Figure 15.7). The aspects of slope stability are discussed in detail in Chapter 9. Before the foundation is laid, underground as well as surface waters have to be diverted altogether, and the loose material—the overburden of rocks—must be removed completely. The foliated, jointed and thinly bedded rocks need to be bolted if possible. It is desirable to have a lining of damp-proof material on the foundation, as well as on drainage ditches behind the walls on the side of a hill slope, to prevent the access of water that accumulates behind or underneath the walls.

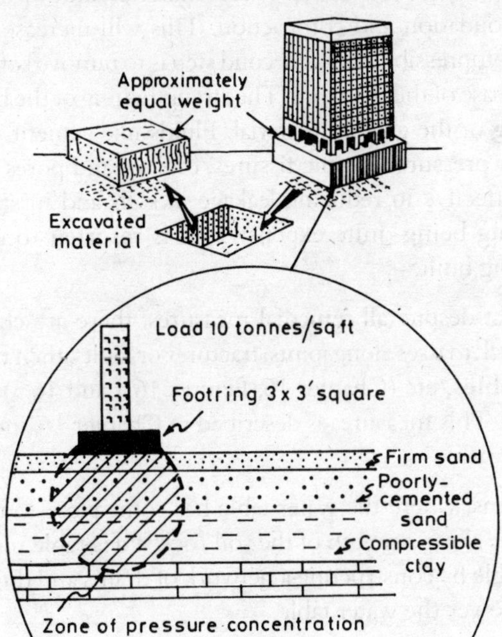


**Figure 15.6.** Soft clayey soil removed to the extent that the amount of excavated material is nearly equal to the estimated load of the structure. A structure built in this manner virtually floats on the soft soil. *Inset* shows distribution of pressure from the superstructure on the ground material. (Modified after Krynine and Judd, 1957).

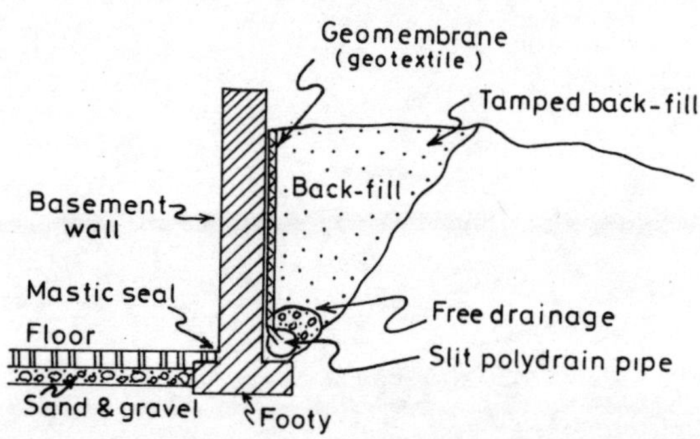


**Figure 15.7.** In hilly terrain, the damp-proof foundation of a building is keyed to the firm ground. Waterproof geomembranes, coupled with perforated pipes, drain out water from behind the wall. The porous permeable bed below the floor serves the same purpose.

## Remedial Measures to Strengthen Bedrock and Soil

Measures to ensure the stability of slopes and strengthening of the ground have been dealt with at length in Chapter 9. The remedies suggested here supplement those measures. To begin, the foundation material should be subjected to the processes of consolidation and compaction. This will increase the shear strength and diminish its compressibility. The second step is to remove soft and loose material from beneath the base of the structure. The strengthening of the building foundation requires **grouting** of the ground material. Fluids like cement, bitumen and resin are injected under pressure into the fissures, cavities and pores of rocks/soils. The process is quite effective in reducing leakage in soils and in strengthening rocks. However, grouting being quite expensive, it is resorted to only when critical structures are being built.

If it is realized that despite all remedial measures, there are chances of failure or displacement of rock masses along joints, fractures or faults, then **rock bolts**, **anchor bars**, **stressed cables**, etc (Chapter 16, Figures 16.4 and 16.5) should be used to reinforce the rock. This measure, as described in Chapter 16, increases the stability of slopes in hill areas.

In waterlogged areas, where the water table has come close to the ground surface and diminished the shear strength of the soil/rock, a desirable measure would be to lower the water table by constructing a network of drains and ditches. The draining out of water will lower the water table.

To overcome the problem of swelling clays, measures are taken to irrigate soils only upto the point of saturation. The application of sufficiently heavy loads counteracts the pressure of swelling. A costly method is chemical treatment—injecting lime, fly ash, etc into the soil. By adding lime to the clay, its plasticity is reduced considerably, so that it becomes resistant to water-softening. Bitumen binds the particles and protects them from the erosive action of water. In short, if the cause of swelling—water—is prevented from reaching the clay, there will be no problem to the foundation.

# 16

# Constructing Transport Routes Without Causing Environmental Distress

## LIFELINES IN THE DEVELOPMENT PARADIGM

Roads and railway lines connect sites of inhabitation, centres of trade and commerce, and areas of production and natural resources—becoming lifelines in the development paradigm of a country. Canals that convey precious water, for various uses, also fall into the same category. Routes of transportation play an important role in ensuring connectivity and convenience—hence, their alignments are most often laid in accordance with the needs of the people, with little regard to the geological conditions and environmental sensitiveness of the terrain. It falls upon geologists to find ways to pursue development, at minimum damage and distress to the environment and communities living along the proposed routes of transportation (Table 16.1).

## THE FOUR ASPECTS OF ROUTE DEVELOPMENT

The first and the most important aspect in route development is to plot the proposed road alignment on a large-scale and up-to-date topographical map. Once the alignment is marked on the map, it becomes possible to circumvent disadvantageous and problematic reaches, and therefore revise the alignment so that it serves the maximum number of habitations, without unnecessarily encroaching upon settlements, agricultural fields and other facilities. The development of routes encompasses four aspects of construction:

(i) Making embankments in the plains,

(ii) Cutting slopes in the hills,

(iii) Passing through hill-barriers by tunnelling, and

(iv) Crossing rivers by building bridges.

In other words, the operations embrace activities in four different situations, and broadly, in two different terrains. It is imperative to ensure that the routes do not pass through geologically unstable and/or ecologically sensitive areas. One can appreciate the dimensions of the problem when one realises that the aggregate length of roads in India is 2,466,000 km (1,072,000 km unmetalled) and railway lines, 68,800 km (Anon, 2001).

**Table 16.1.** Evaluation of Environmental Impact of Road-building.

A. *General Setting of Region*
   1. Topography
   2. Geology
   3. Climatic condition
   4. Existing network of roads and paths
   5. Route alignment of the proposed road

B. *Objectives of the Road Project*
   1. Size of the area and population it will serve
   2. Volume of traffic expected in the first, second and third years

C. *Technical Features*
   1. Length of road
   2. Width and type of road
   3. Slopes of hills it will cut through, and heights of road-cut

D. *State of the Environment of the Belt through which the Road will Pass*
   1. Hillside stability and problem of erosion
   2. Seismicity in the region
   3. Identification of hazard-prone zones
   4. Faunal communities
   5. Flora—extent and types
   6. Parts of road prone to flood-hazards

E. *Impact of Road Construction on the Environment*
   1. On stability of hillsides
   2. On hydrological conditions—surface and underground water
   3. On forest ecosystems
   4. On farms and settlements

F. *Measures Proposed to Mitigate/Minimize Adverse Impacts*
   1. Balancing cut and fill
   2. Disposing of debris
   3. Reclaiming pits and securing embankments
   4. Preventing mass-movements
   5. Containing blasting
   6. Landscaping route belt

G. *Socio-economic Concerns*
   1. Employment potential
   2. Displacement and/or flight of people

The factors that need to be taken into consideration are:

(i) The stability of the ground and of the hillslope to be cut.

(ii) The shear strength and bearing capacity of the materials that form the ground and the slope.

(iii) The groundwater condition, particularly the level of the water table and the pattern of flow.

(iv) The natural hazards that are likely to occur along the route, particularly where installations of vital importance are to be located.

## Making Road Embankments

In plains or flatter terrains of plateaus, road-pavements and railway lines are emplaced on **rock-fill** or **earthern embankments** (Figure 16.1). A crucial consideration during construction is the possibility of unequal or differential settlement of the foundation material—the ground beneath the embankments. This happens if the ground material is **incompetent**—soft or yielding. Materials like clay and mud, and even silt, peat, landfill and loose gravel do not have the required bearing capacity to withstand the load of heavy traffic on roads or railway lines. Successions of alternating competent and incompetent material are likewise prone to differential settlement, especially where water enters the incompetent layers (Figure 16.1A). The incompetent material yields to plastic deformation. This manifests itself in the bulging, wrinkling and spreading of the road. The subject has been dealt with in some detail in Chapter 15.

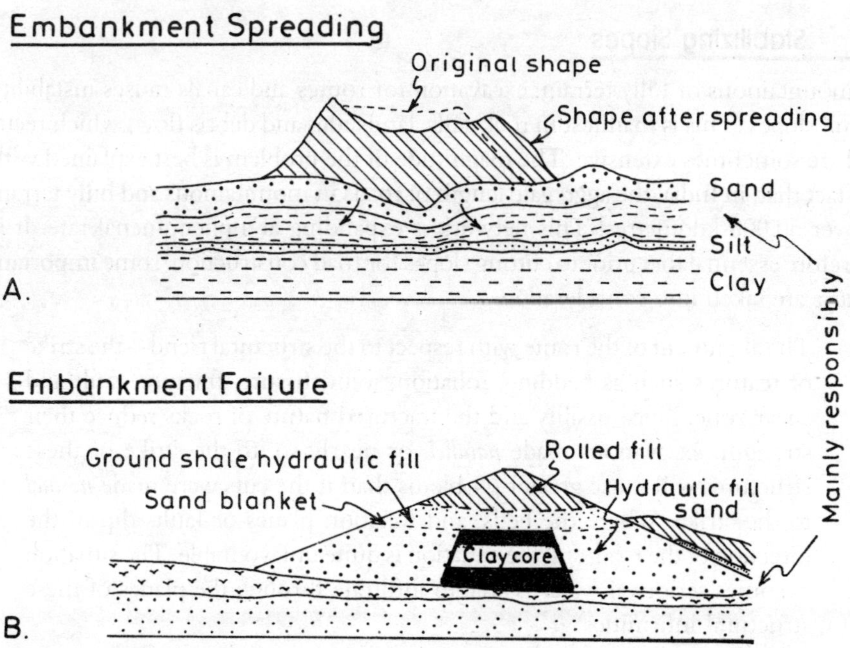


**Figure 16.1.** (A) Alternating competent and incompetent material beneath the road pavement may cause failure of the embankment due to the incompetent layer that yields to plastic deformation.

(B) The earthen core of a rock-fill embankment—made to prevent seepage across the structure—may fail when shaken severely by an earthquake, or due to waterlogging of the ground.

In view of these problems, it is desirable to construct effective drainage systems that intercept water, and remove it from the embankment as well as from the ground below it. It is imperative to keep the water table well below the ground surface under embankments. This can be done by constructing longitudinal ditches and gullies and linking these with culverts. Preventing the entry of water will not only control denudation of the embankment, but also forestall **piping** and **tunnel erosion** that, in many cases, lead to the collapse of the road.

In the making of roads, adequately compacted embankment material is covered with multi-layered pavements. The pavement must comprise of three layers: (i) substrate of compacted gravel, (ii) coarse-aggregate made of gravel, granule and coarse sand, mixed with asphalt and laid to the definite thickness (**base coat**), and (iii) fine sand mixed with asphalt, laid to a thickness of 5–10 mm and pressed by road rollers to form a **seal coat**. It is this seal coat that protects the exposed coarse-aggregate layer from the ravages of rain and rolling stocks, ensures longer life to the road, and also reduces wear and tear of vehicular tyres. Unfortunately, most Indian roads are bereft of the seal coat and therefore prone to very early damage. It is difficult to estimate how much of the 1,394,000 km of metalled roads in India have a protective seal coat.

## Stabilizing Slopes

In mountainous or hilly terrain, excavation for routes and canals causes instability of hill slopes. This is manifest in rock-falls, landslides and debris flows which recur and are sometimes extensive. The magnitude of the problem is best explained with the fact that, in India, the aggregate length of roads in mountainous and hilly terrain is over 50,000 kilometres! This network is expanding at a phenomenal rate. It is therefore essential that prior to cutting slopes for road construction, some important factors are taken into consideration.

(i) The alignment of the route with respect to the structural trend—the strike of features such as bedding, foliation, joint, fissure, fracture, fault and shear zone. Since fissility and the fractured nature of rocks reduce their strength, excavations made *parallel*, or nearly so, to the strike of these structures will cause greater problems than if the cuts were made *normal* to the strike. Where the beds, and foliation planes or faults dip in the direction of the road-cut, the situation is quite unfavourable. The situation becomes serious if water is able to infiltrate through the planes of these structural infirmities.

(ii) The composition of rocks and/or soils excavated. Compact, dense, consolidated or **competent** rocks/soils cause less problems on the slope of the road-cut than soft, loose, yielding or **incompetent** material. Alternate layers of competent and incompetent material are quite treacherous in the matter of slope stability.

(iii) The height and steepness of the road-cut. These depend on the shear strengths of the slope material, on rocks/soils. The steeper the slope, the greater would be the chances of slope failure. If the rocks are composed of slate, phyllite, schist and gneiss, or unconsolidated to semi-consolidated material, the slope-cut should be gentle—between ½ : 1 and ¼ : 1 (Figure 16.2). Slopes higher than these values are likely to be affected by slips and slides. The average values of slopes are as follows:

— granites, basalts, rhyolites, gneisses, schists, marbles, massive sandstones and dolomites ¼ : 1 to ½ : 1

— slates, phyllites, interbedded sandstones, shales and limestones ½ : 1 to ¾ : 1

— claystones, mudstones and siltstones ¾ : 1 to 1 : 1

It is advisable to bear in mind the implications of cutting slopes for road-cuts with a steepness that is more or less than the **angle of repose**. The angle of repose is the steepest angle at which the sloping surface remains stable. If stability is to be ensured, it is advisable to avoid talus-fans and deposits of landslides and debris flows. If that is not possible, it would be safer to cut the road at the upper part of the debris fans or cones, or at concave lower ends, where bigger clasts lie in a relatively stable disposition. Likewise, fluvial terraces made up of unconsolidated or semi-consolidated gravel are as unstable as debris deposits for making road-cuts.

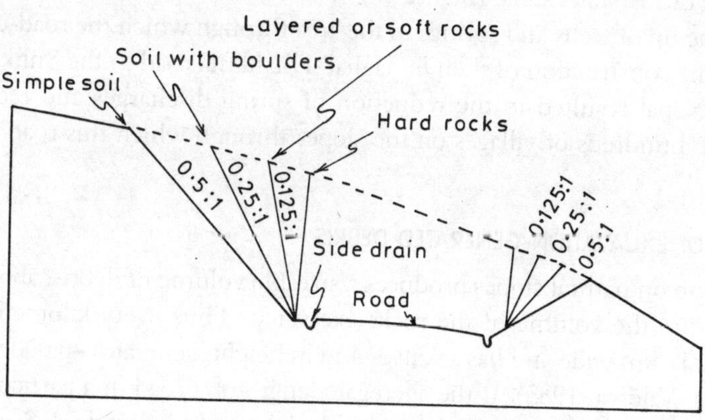


**Figure 16.2.** The angle of repose—the steepest angle at which the sloping surface remains stable—depends on the nature of the material making the slope. The height of the slope-cut is another factor that has bearing on slope stability.

If an unstable belt like a shear zone, fault zone, or highly jointed rock comes in the way, it would be safer to keep the height and the slope of the road-cut to the minimum. This can be done by breaking the wall of the cut and developing benches (1.5–2-m wide for every 10–15-m elevation above the road), and by lessening the load on the slope. The construction of wide shallow ditches (or drains) on the natural slope helps the slope retain its stability (Figure 16.3). The drain should be led into culverts that have an original-bed gradient so that there is no accumulation of water and deposition of debris in the channel.

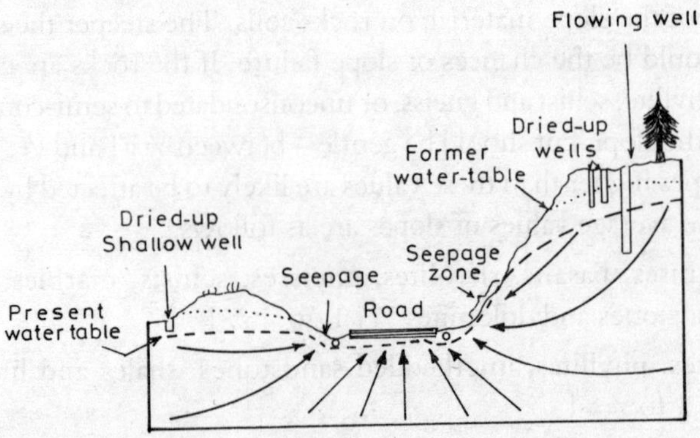


**Figure 16.3.** Cutting of hill slopes causes interception of the water table. This manifests itself in the origin of springs and seepages on the wall of the cut. The ultimate consequence is the drying up of wells and springs.

## GROUNDWATER INTERCEPTION

Since the water table follows the configuration of the topography, the cutting of natural hill slopes in many cases results in the interception of groundwater. This may bring about a disruption in underground circulation. The intercepted groundwater oozes out in the form of seepages, or spouts out as springs, along the walls of road-cuts. This culminates in the beheading of the aquifer and the resultant lowering of the water table (Figure 16.3). The lowering of the water table leads to the drying up of wells and springs in the area through which the road-cut has been made. The construction of the Dholaghat–Dhankuta road in the Sunkosi valley in eastern Nepal resulted in the reduction of spring discharges and reduced water supply to hundreds of villages on the slopes through which this road passed (Jha, 1982).

## EFFECTS OF EXCAVATION-GENERATED DEBRIS

Excavation on natural slopes produces a sizeable volume of debris, about one-and-a-half times the volume of the rocks excavated. Thus every kilometre stretch of road that is 4 m wide and has a wall 3–4 m in height, generates 40,000 to 60,000 $m^3$ of debris (Valdiya, 1985). If the aggregate length of a road in a particular region is 100 km, the total amount of debris produced would be up to 4–8 million cubic metres! Now, in the mountainous domains of India, the aggregate length of the road is more than 50,000 km. Which means that the volume of debris produced from road construction would have been 200,000 to 400,000 million cubic metres!

This waste material, when allowed to slide down the slopes, kills forest vegetation, pastures and agricultural fields, buries springs, and chokes streams that lie below the road. And this is not all. Every kilometre of constructed road in the Himalaya, on average, generates debris at the rate of 550 $m^3$ per year—due to rock-falls, slips and slumps during the monsoon (Valdiya, 1985). According to one estimate, nearly 57,700 ha of valuable forests (1.6% of the total forest cover in India) were lost in the twenty-two years between 1951–73 due to construction of roads (Sharma, 1977).

## DISPOSAL OF DEBRIS

The generation of debris is inevitable when slopes are cut for making roads, railway tracks and canals. The amount of waste produced is very large and needs to be disposed of appropriately. This can be done in a variety of ways. Depending on its usability, the excavated material can be used to build parapet walls along the length of the road, dumped behind the retaining wall especially in concave bends of the road (Figures 9.12A and 16.4A), utilized for developing roadside parking spots and gardens, and hauled out as building material.

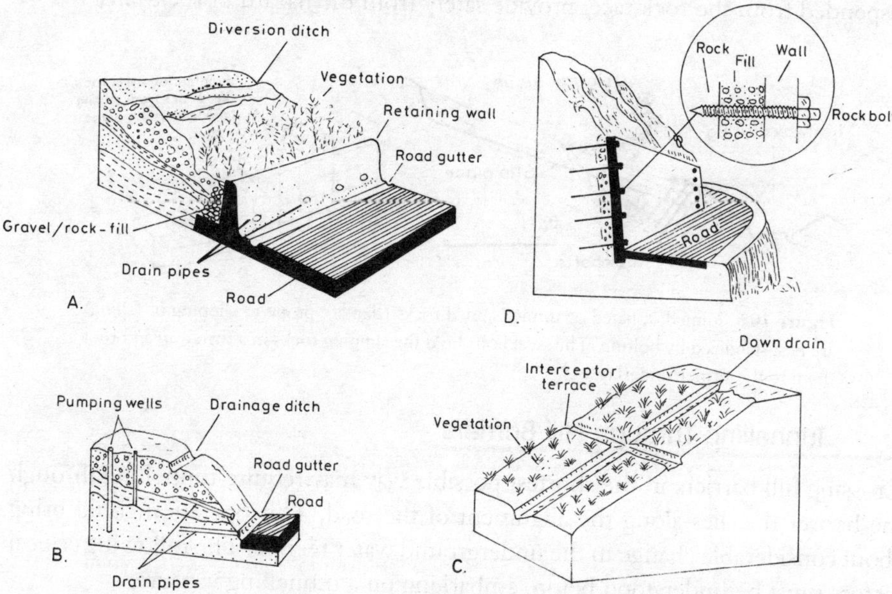


**Figure 16.4.** (A) Retaining wall holds back the loose material of the slope and prevents it from sliding. Diversion ditch above the road, on the slope, stops water from its act of destabilization. Vegetal carpet puts a protective shield on the slope.

(B) Drainage ditch above the road-cut, on the slope, prevents runoff water from entering vulnerable spots. Vertical pipes with pumps and inclined perforated pipes drain out water from the water-saturated loose material.

(C) Putting a protective cover of vegetation and constructing a drainage network stabilizes the slope.

(D) Retaining wall is securely bolted to the rocks of the slope. *Inset* shows the nature of the bolt.

## ENGINEERING STRUCTURES TO STABILIZE SLOPES

The first and very crucial step to control slope failure in road-cuts is to construct **drainage ditches**, in order to divert runoff water from the slope above the road-cut (Figure 16.4A and Plate 5D). Treating the slope with gunite, bitumen or clay will prevent underground percolation of rainwater, into the loose mass of soil or into vulnerable rocks. Simultaneously, an operation to build a vegetal carpet on the slope should be initiated (Figure 16.4C). Plants that have deeper roots and thicker foliage are preferred.

The second important step is building a **breast wall** (Figures 9.12, 16.4A, 16.4D) or a **buttress wall**. If the rocks of the slope are fractured, highly jointed or fissile, it is best to bolt the retaining wall to the rocks by driving in **rock bolts** (Figures 16.4D and 16.5). The type and pattern of rock bolts are decided on the basis of the geometry of divisional planes in the rocks and the quality or nature of the rock itself. In thinly bedded or foliated rocks, ordinary rock bolts may not prove effective, and grouted bolts may have to be used (G. Pant, per. com. 2002). The deeper the steel bolts, the greater would be the strengthening of the slope. Nets of steel-wire, suspended from the rock face, provide safety from the hazard of rock-falls.

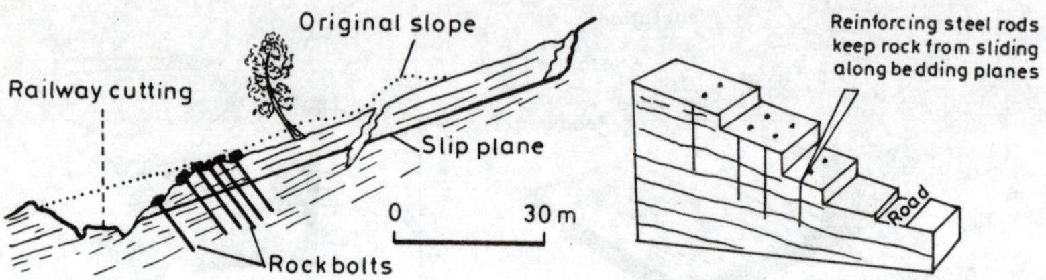


**Figure 16.5.** Jointed, foliated or thinly bedded rocks (that are prone to slipping or falling) are best stabilized by bolting. The steel bolts hold the slipping rocks in a firm grip and root them to the firmer underlayers.

## Tunnelling Through Hill Barriers

Crossing hill barriers in the shortest possible way may require tunnelling through the barrier that lies along the alignment of the road. This operation could bring about considerable change in the underground water regime. The following critical factors must be understood before embarking on a tunnelling venture:

(i) Structural set-up and lithological making of the hill barrier.

(ii) Measure of competence, shear strength and bearing capacity of the rocks and/or soils along the tunnel.

(iii) Stress condition inside the tunnel. It is this stress that brings about deformation and displacement of the tunnel lining and structures.

(iv) Position of the water table and patterns of groundwater flow that may cause problems like seepage and gushout.

(v) Hardness and workability of the rocks along the tunnel, so that engineers have a fair idea of the problems likely to arise during operations.

It is obvious that a comprehensive geotechnical investigation is a pre-requisite before embarking on tunnel projects.

### TUNNEL ALIGNMENT AND STRUCTURAL DISCONTINUITIES

Bedding planes, cleavages, joints, fissures, fractures and faults are structural discontinuities in rocks. These are planes of weakness that *must* be taken into consideration before and during tunnelling. If these structures are horizontal, or nearly so, and widely spaced there will be fewer chances of failure in the tunnel, on account of the **bridging effect** in the discrete blocks. The innate self-balancing shearing strength of rocks is then called into play.

If the tunnel is driven parallel to the strike of the bedding (Figure 16.6A), whether vertical or inclined, the situation would cause instability—even if the rock above is held up by frictional resistance on the bedding plane. If the tunnel were driven entirely through one rock, the weight of the rock would be brought to bear on the tunnel lining and there would be seepage through the permeable layers during the construction (Figure 16.6A). However, if the tunnel is cut partly through harder rocks and partly through softer, there is the danger of sliding or slipping along the contact planes. The 339-m long Jaldhaka tunnel in the Darjiling hills, driven parallel to the strike of the Daling–Darjiling metamorphic assemblage, experienced frequent roof collapse and heavy seepage through the sheared quartzite, gneiss and phyllite, and along the faults.

In the case of synclines (Figure 16.6A), where high compression prevails in the core areas, the joints and fracture planes converge upwards so that wedge-shaped blocks are liable to drop down (Figure 16.6C), even as percolating water seeps freely into the tunnel during construction (Figure 16.6A). This kind of problem was faced in the 6.6-km long Loktak tunnel in Manipur, which traverses a N–S trending folded hill made up of the Disang–Barail succession of sandstone alternating with shale. Shear zones and open joints serve as passages for underground water causing intractable problems inside tunnels (Figure 16.6B). However, if a tunnel follows the anticlinal core axis (Figure 16.6D), the problems of rock-fall, sliding and seepage are uncommon.

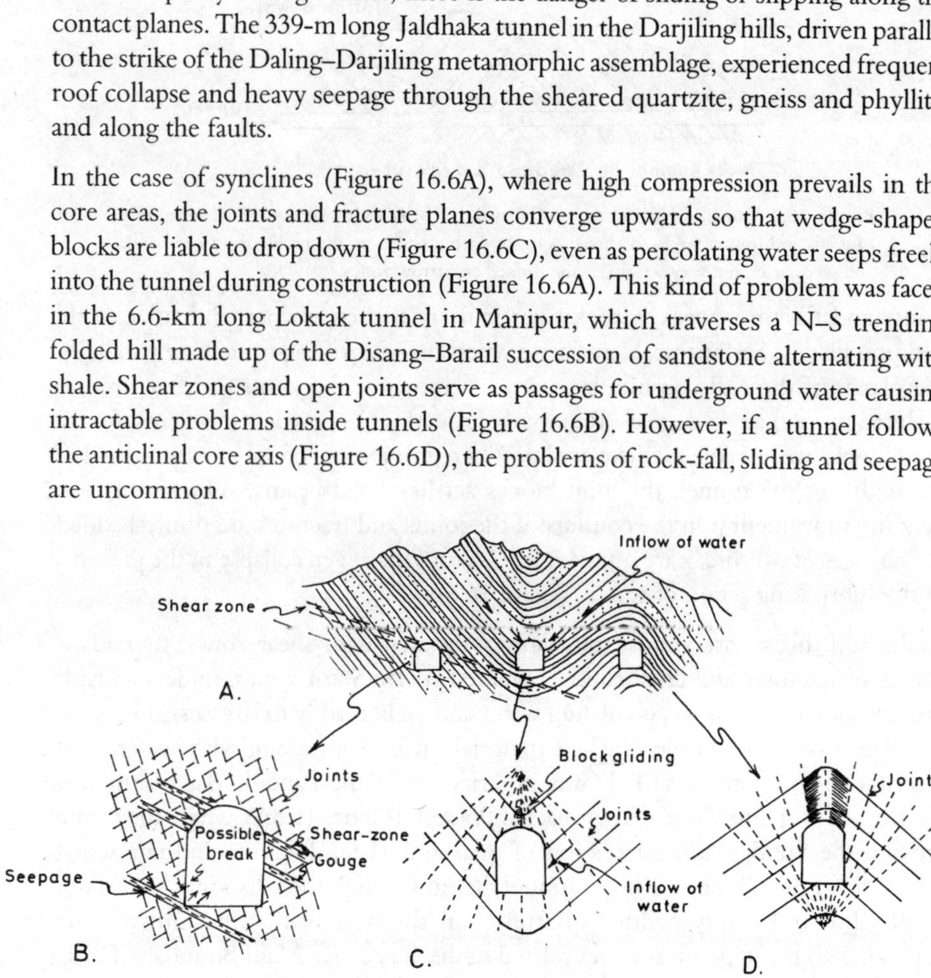


**Figure 16.6.** (A) Tunnel driven parallel to the strike of the beds of the folded succession of permeable (dotted) and impermeable (lined) rocks. The permeable beds allow percolation of water.

(B) Shear zones and open joints serve as passages of underground water.

(C) In the synclinal core, the joints converge upwards, permitting rock blocks to drop down and the underground water to flow freely into the tunnel.

(D) In the anticlinal core, the situation is entirely opposite.

If a tunnel cuts through the rocks normal to their strikes (Figure 16.7) and if the beds are steeply dipping or vertical, the probability of **overbreak** and seepage is not high. If they do occur, their extent will be limited to shear zones and other vulnerable spots. The Ramganga tunnel at Kalagarh (Uttaranchal) is driven normal to the strike of the Siwalik succession of sandstone, mudstone and shale. It faced the problem of overbreak, tunnel-wall failures and seepage only in the narrow zones of active thrusts and faults.

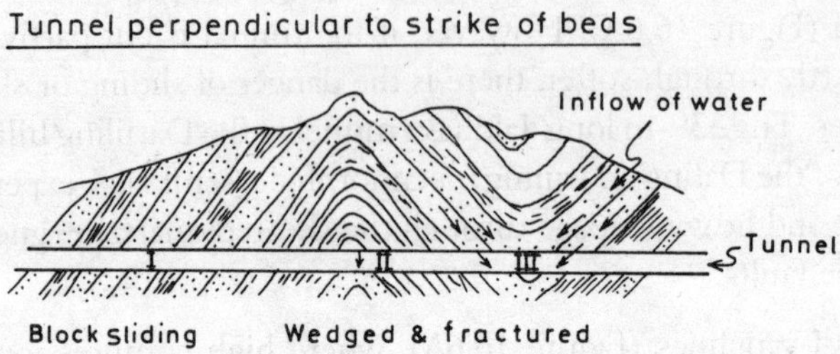


**Figure 16.7.** Tunnel driven normal to the strike of beds with a succession of permeable (dotted) and impermeable (dashed) rocks. In such a situation, the problems of overbreak and seepage are not very serious, being limited to narrow zones.

Joints and fractures weaken rocks, particularly if they are perpendicular to the bedding planes. Serious problems arise when the axes of the tunnel are parallel to the planes of these discontinuities, particularly where the joints and fractures are inclined more than 45 degrees. Randomly jointed rocks pose less problems, for the joint blocks are not able to move or slip. Where the spacing of the joint is more than the width of the tunnel, the joint blocks act like solid beams, subject only to a bending movement. On the contrary, if the joints and fractures are thinly-bedded, the chances of overbreak are quite high. The roof may even collapse in the presence of the lubricating action of water (Figure 16.6).

Faults and thrusts are almost invariably associated with shear zones, myriads of planes of fractures and crushed or powdered rocks. Shear zones, understandably, provide easy pathways to percolating water and yield readily to stresses and strains. In some cases, the whole crushed material gushes out along with water, with destructive consequences. This was experienced in the Pandoh–Baggi tunnel of the Beas–Satluj Link Project in Himachal Pradesh (Figure 16.8A), where the tunnel crossed the Sundernagar succession of quartzite, slate, dolomite and talc schist. Likewise, in the Chhibro–Khodri tunnel driven through the Chandpur succession of phyllite and thinly-bedded quartzite (in the Tons valley in southwestern Uttaranchal) the wide shear zones related to the very active Main Boundary Thrust (MBT) and Nahan Thrust (NT) were the cause of extremely difficult problems; particularly in the zone of crushed rocks (belonging to the Subathu Formation) that lie sandwiched between the two fault planes (Figure 16.8B). To overcome these problems, the alignment of the tunnel was given a kink, making the tunnel traverse only 250 m of the thrust-bound zone, instead of the original 700–750-m stretch in a straight alignment. The width of the tunnel was also increased to nine metres from the original seven metres (Shome, 1983).

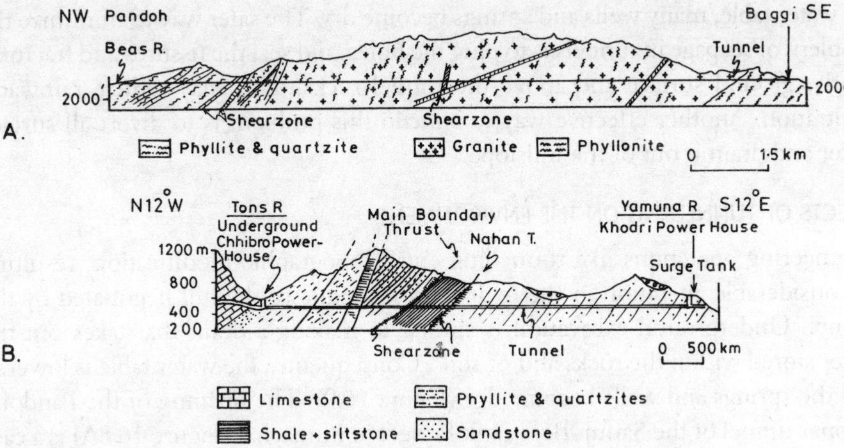


**Figure 16.8.** Sheared, shattered and crushed rocks associated with active thrusts and faults pose serious problems of overbreaks, roof collapses and violent gushouts.

(A) The Beas–Satluj Link Project along the Pandoh–Baggi tunnel

(B) The Chhibro–Khodri tunnel in the Tons valley (After Shome, 1983).

The Uri Project tunnel through the Pir Panjal range in Kashmir encountered serious problems of squeezing of rocks as it crossed the Panjal Thrust zone, characterized by deformation-prone sedimentary and metamorphic rocks, principally of the Eocene age. Spiling and multiple drift methods were used to give sufficient stand-up time in order to install proper rock support and reinforcement (Stille et al, 1998). The 27-km long tunnel of the Nathpa–Jhakri Project in Kinnaur (Himachal), ensuring a drop of 450 m in the Satluj river, cuts through a 500-m shear zone near Ratanpur and a fault near Dajkhas, in a succession of high-grade metamorphic rocks. The shear zone created a serious distortion in the tunnel supporting system, necessitating multi-drift tunnelling methods, coupled with drainage and rock reinforcement (Gupta, 1998). The faults brought in hot (40°–60°C) water at the rate of 100 litres/sec, and there were massive rock slides in the river valley.

During the construction stage, the tunnel connecting the reservoirs of the Kundah Project in Tamilnadu was flooded with a sudden rush of water, mixed with sediments, about 30 m below the fault depression. Obviously, the water gushed out through the shear zone, causing havoc. The fault gouge comprised swelling clays also. The presence of such clays may block the flow of interstitial water, leading to the build-up of pore-water pressure. When the pore-water pressure exceeds the threshold pressure, the impounded water bursts out and floods the tunnels.

## REMEDIAL MEASURES CONCERNING TUNNELS

Leaking and gushing out of water in tunnels can be lessened by lowering the water table to below the level of the tunnel, before tunnelling starts. This is done by drilling drainage holes from the head of the tunnel, or drilling deep wells and pumping out the water. This is an expensive and time-consuming process. Moreover, this is counter-productive for inhabitants of the hillslope—with the lowering of the water table, many wells and springs become dry. The safer way to minimize the problem of seepage in tunnels is to plug the holes, and seal the fissures and fractures on the ground surface and above-the-tunnel rocks. This may reduce rainwater infiltration. Another effective way to contain this problem is to divert all surface water and drain it out of the hillslope.

## EFFECTS OF TUNNELLING ON THE ENVIRONMENT

Engineering operations like tunnelling entail topographic modification, resulting in considerable alteration in the hydrological regime of the hill negotiated by the tunnel. Underground excavation is similar to making a drain that takes out the water stored within the rocks and/or soils. Consequently, the water table is lowered and the springs and wells become dry (Figure 16.9). The opening of the Pandoh–Slapper tunnel of the Satluj–Beas Link Project in Himachal (Figure 16.8A) is a case in example. Nearly 50% of springs in the areas lost their discharges as a consequence. The extent and rate of draining out of water in this manner depends on the permeability of the rocks—their ability to allow flow of water through interconnected holes, fissures, fractures and faults. As stated above, water commonly rushes out with considerable force through shear zones and related faults, as witnessed in the Banihal tunnel that runs through the Pir Panjal range in Kashmir.

In view of these problems, it is desirable to carry out a detailed Environmental Impact Assessment, simultaneous with detailed geotechnical investigations, before the start of a tunnel project.

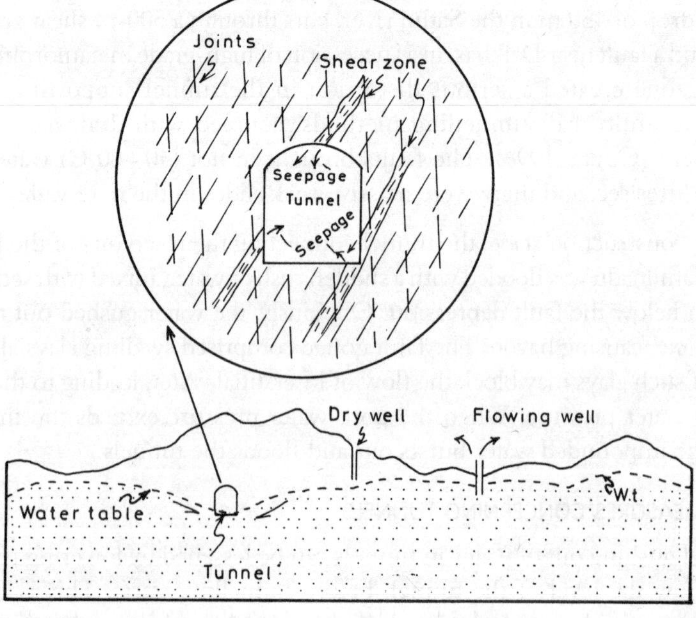


**Figure 16.9.** A tunnel serves as an underground drain, taking out groundwater and lowering the water table. *Inset* shows shear zones and joints that provide ready pathways to interstitial water.

## MAKING AN UNDERGROUND PASSAGE

The Kolkata Metro Railway runs through an underground tunnel, drilled through a thick succession of unconsolidated sand, silt and clay, with localized pockets of organic matter representing former swamps. The entry of water from the Hugli river keeps the sediment perennially moist. Large sand bodies within this succession bring volumes of underground water into the tunnel. Even discontinuous isolated pockets of sand, with clay, drain out small quantities of water into the tunnel. The consequent hydration of clays prone to swelling exerted pressure on the tunnel lining. The organic matter present within the sediment aggravated the problem of instability, often culminating in failure. The rapid development of buildings near the belt of the tunnel alignment continues to exert pressure on the tunnel structure.

In view of such problems it is essential to provide engineering support at a very early stage of the operation. In order to dewater the clayey sediments, the ejector system that creates a vacuum around the tunnel and draws out water, can be used. This water can then be pumped out of the tunnel.

## Building Bridges Across Rivers

Bridges are built to create routes across streams and rivers. There is, therefore, limited choice in selecting sites for them. The geologist's job is merely to assess the strength of the rocks/soils that would provide terminal and intermediate support to the bridges. In other words, it is important to know how the rocks/soils would react to the abutments and piers when rolling stocks move through the bridges.

Selection of the type of bridge is governed by the geological conditions of the ground, and the dynamics of flowing water in the channel during its swiftest phase—the flood. The bridge must remain safe and secure even during: (i) peak discharges, (ii) eddy actions, (iii) landslides and/or rock-falls, and (iv) foundation failures, whatever the cause.

If the channel is not wide and the abutment rock/soil is firm and stable and possesses the ability to withstand the combined loads of the bridge structure and vehicular traffic, a **simple beam bridge** (Figure 5.10A) is the obvious choice. However, if the channel is wide enough to require intermediate support (piers), that could stand on the firm foundation of the riverbed, a **cantilever bridge** is constructed (Figure 16.10B). On the other hand if the abutment is located on a poor foundation, it is the piers that have to take the load. They must therefore reach the firm base deeper down to cope with the problem of differential settlement, due to the presence of soft yielding (incompetent) material in the foundation (Chapter 15). For example, the Dholpur bridge on the Chambal river was constructed on subrecent soft conglomerate, which could not bear the load of the structure and rolling stocks and consequently failed. The remedy lay in distributing the load over a larger area of the sensitive spot, by making the foundation wide. The **pile foundation** laterally distributes the pressure on the soft soil and/or sediment (Figure 15.4C). This is yet another method of overcoming the hazard of differential settlement.

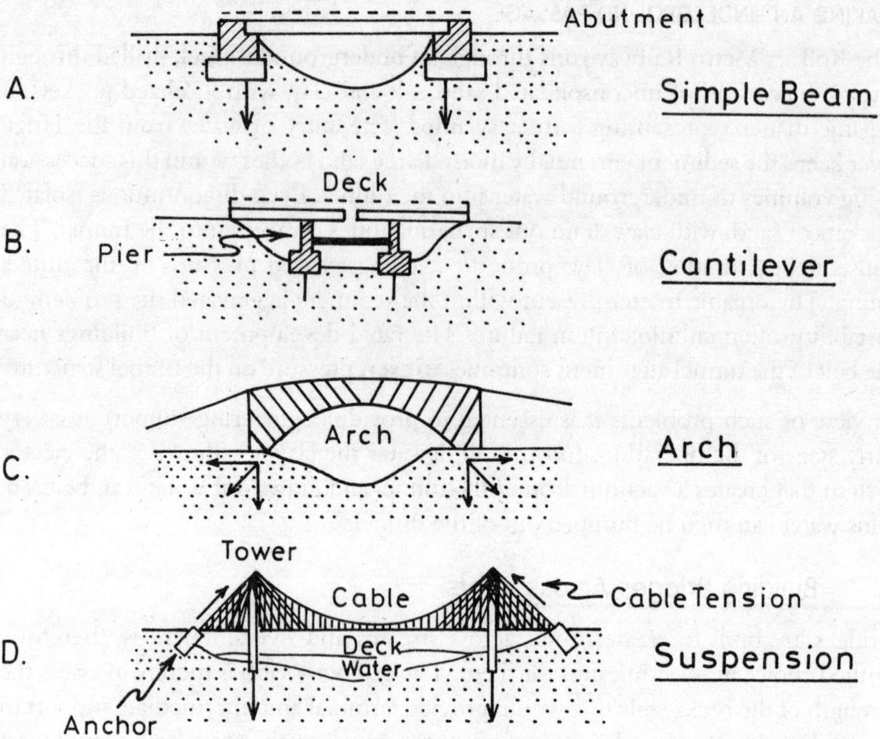


**Figure 16.10.** Selection of bridge-type depends onthe geological condition of the foundation material and the dynamics of streamwater flowing through the channel, particularly when the flow is in the swiftest phase (Based on Leverson, 1980).

In addition to vertical loads, there is always a horizontal thrust or lateral pressure on the abutment. In such cases, the **arch bridge** or **rigid-frame bridge** (Figure 16.10C) is built. These are able to withstand stresses suitably. In a **suspension bridge** (Figure 16.10D), the superstructure is anchored with the help of cables to rocks or concrete blocks on abutments that are sound and stable. This type of bridge is well-suited to mountains, where the valleys are narrow, the slopes are rocky and the streamflow very swift.

# 17

# Engineering Water Resources Development Without Causing Distress

## Reservoirs: An Artificial Creation

Storing water—for irrigation, electricity generation and flood moderation, etc—requires building dams across rivers. There is no denying that the benefits of dams are many. Farmers are assured a regulated supply of water and enjoy the resultant freedom from the vagaries of unpredictable weather. An estimate shows that the total output of food production increases by about 10% because of irrigation from reservoirs behind big dams. Hydro-electric power generated by dams supplies the growing demands of cities, towns and industries. And, where reservoirs are kept half or partially empty before the onset of heavy rains, it is even possible to moderate the fury of floods.

However, the fact remains that these artificial creations (reservoirs and dams) force considerable changes upon the environment. Flowing-water ecosystems are replaced by standing-water regimes. This transformation brings about a great deal of change in the landscape, and the life of people who are favourably or adversely affected by these changes. The impact of the dam is felt in the river basin, both upstream and downstream of the blockage. It is therefore necessary that geotechnical investigations confirm the soundness of the site, and conduct an appraisal of any likely environmental impact, before launching a water resources development project. It is imperative that the planners, the developers, and the people know the benefits as well as the fallouts resulting from the effort to tame a river. In other words, technical feasibility and economic gains should not be the only consideration while building dams. Favourable assessment in terms of environmental changes and their socio-economic implications are important criteria before constructing an artificial structure across rivers.

## Dam Sites and Dam Types

The hydrological and geological conditions of river basins govern the selection of sites and types of dams. A dam should be built in harmony with the natural environment, and be strong enough to face the geodynamic conditions of the region. Ideally, a dam should be located in a valley where: (i) the discharge is sufficient to meet the anticipated requirements, (ii) it is narrowest, so that less construction material is required, and (iii) the foundation (riverbed) is strong enough to bear the load of the structure. A factor of considerable importance is the availability of construction material close to the construction site.

If the valley is very narrow and the exposed rocks of the steep valley slopes are strong enough to take the load, an **arch dam** (Figure 17.1A) is the appropriate structure. Since all stresses are transmitted onto the abutments, there should be no weakness parallel to the direction of the arch stresses. The topographical contours downstream of the dam should not show recession from the river (G. Pant, per.com., 2002). An example is the 169-m high Idukki dam across the Periyar river in the southern Sahyadri.

If the foundation rocks have variable strength and are located in a wide valley, a **buttress dam** (Figure 17.1B) would be the best option. In buttress dams, the sloping decks bear the load of impounded water and accumulated sediment. Vertical buttresses, aligned normal to the plane of these slabs, transmit the pressure to the foundation which must not be weak at those spots.

A **gravity dam** (Figure 17.1C) is suitable for most conditions, owing to the stability provided by its great bulk or weight. Since the entire weight of the dam is transmitted directly onto the foundation, the rocks along the riverbed must be strong enough to take the entire load. The 226-m high Bhakra dam across the Satluj river (in the sandstone–shale succession of the Siwalik terrane of Himachal); the 104-m Koyna dam erected on a foundation of Deccan lavas in Maharashtra; and the 125-m Nagarjunsagar dam across the Krishna river in the terrane of the gneisses of Andhra Pradesh, are all examples of gravity dams.

Where the foundation consists of an alternation of strong and weak rocks, and/or where the possibility of earthquakes is high, a **rock-fill** or **earth-fill dam** is preferred (Figure 17.1D). The 260-m high Tehri dam in the Bhagirathi river is a rock-fill dam built on a succession of phyllite and quartzite. The riverbed is characterized by a multiplicity of shear zones, where there is a very high likelihood of large earthquakes occurring. Another example of the earth-fill dam is the 126-m high Kalagarh dam across the Ramganga river in the Siwalik terrane of sandstone and shale.

If the valley is very wide and embraces variable foundation conditions, a **composite dam** is preferred. It is partly a concrete or masonry dam and partly a rock-fill/earth-fill structure. The Mulla Periyar dam, built in the nineteenth century across the Periyar river in Kerala, is a classic example of a composite gravity dam made of stone masonry and lime concrete. Recently this has been strengthened with reinforced concrete in the downstream section. The 26-km long, 59-m high Hirakud dam across the Mahanadi river in Orissa, flowing through a terrane of granite, quartzite and slate, is an example of a composite dam. The 62-m Bhavani dam, founded on gneiss, amphibolite and schist in Tamilnadu, is another example of this type of dam.

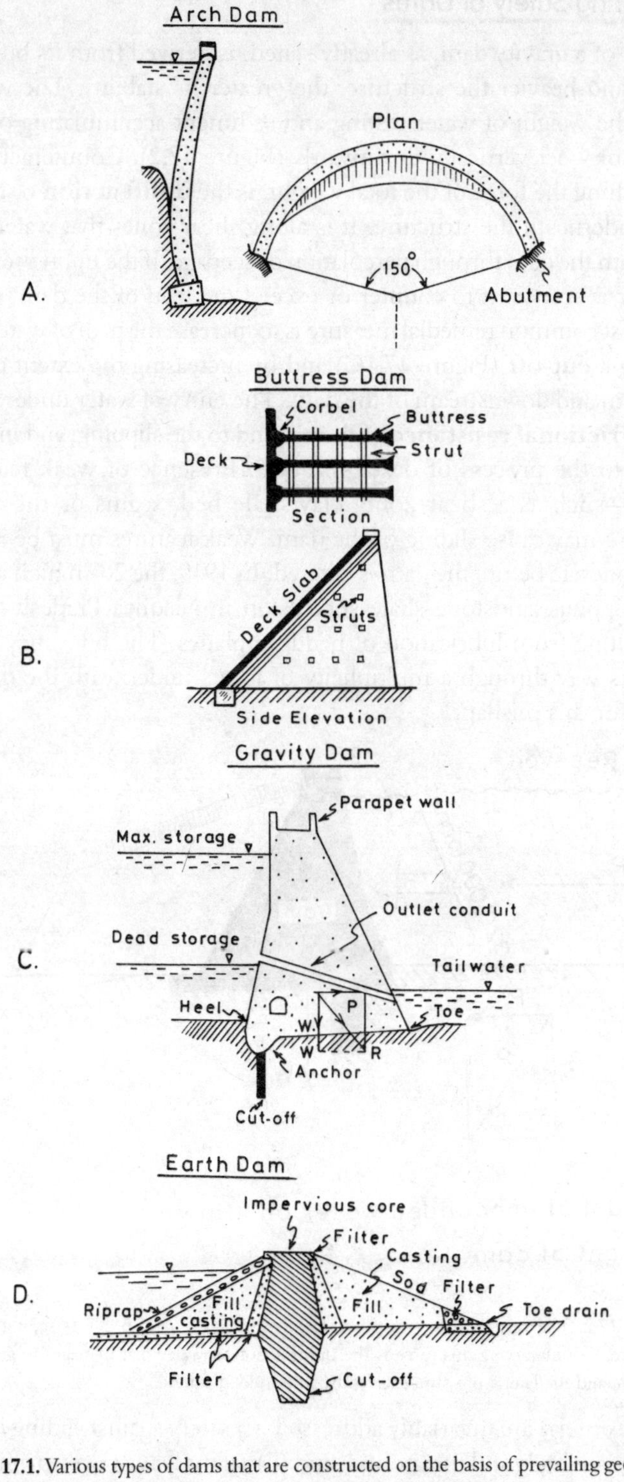


**Figure 17.1.** Various types of dams that are constructed on the basis of prevailing geological and hydrological conditions. (Based on Krynine and Judd, 1957)

## Ensuring Safety of Dams

The stability of a gravity dam, as already stated, is derived from its bulk or weight. The bigger and heavier the structure, the greater its stability. The weight of the dam—plus the weight of water resting and sediment accumulating on its sloping upstream flank—act vertically downwards (Figure 17.2). Counteracting this, and thus diminishing the force of the total weight, is the **uplift action** or **buoyancy** of the water underneath the structure. It is along shear zones that water manages to get underneath the dam through percolation or seepage. If the uplift pressure exerted by this water is sufficient to counter or exceed the load of the dam, the structure fails. The most common remedial measure is to increase the path of water percolation by providing a **cut-off** (Figure 17.1C) and by increasing the extent of the apron, both upstream and downstream of the dam. The entry of water under the dam also reduces the **frictional resistance** of the ground to the slipping, and in this manner contributes to the process of destabilization. Presence of weak features in the foundation—such as a shear zone, clay/shale bed, seams or the bed dipping downstream—may cause sliding of the dam. Weak features must be removed and filled with concrete before the dam is erected. In 1919, the 26-m high Tigra dam on the gently dipping sandstone-shale succession in Madhya Pradesh failed due to sliding, resulting from lubrication of bedding planes. The buoyancy of the water had found its way through a multiplicity of joints underneath the dam, and was responsible for this mishap.

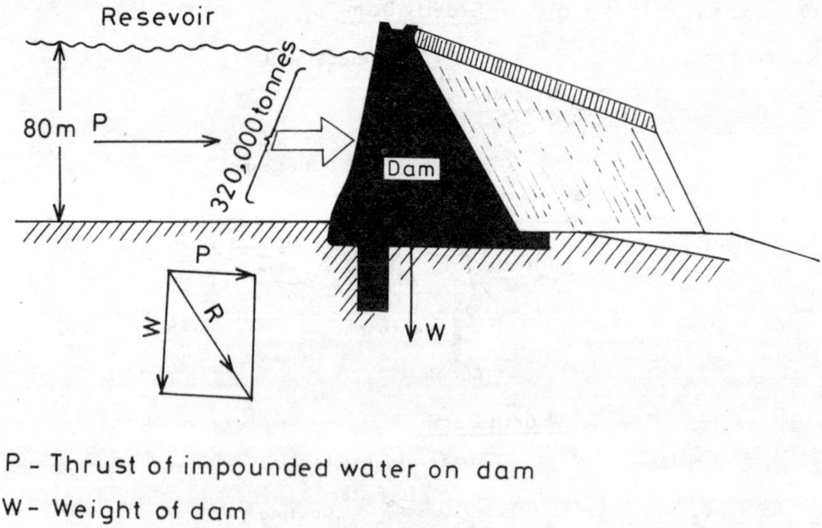


**Figure 17.2.** The vertically downward-acting weight of the dam provides stability to the structure. Counteracting this force is the uplift action of water that manages to get under the dam, and the lateral pressure exerted by impounded water.

Two stability criteria are invariably addressed: (i) safety against sliding *versus* friction between the dam body and the ground, and (ii) safety against overturning in relation to the balance of forces, so that a gravity dam does not overturn due to lateral hydrodynamic forces. This brings out the importance of the geometry of the dam.

Impounded water, exerting strong horizontal pressure, is the biggest destabilizer. The dam topples if and when this pressure exceeds a critical limit. This can happen when seismic shocks and/or powerful storms generate strong surging waves in the impounded water. Similarly, when floodwater **overtops** the dam and the vertical weight felt at the bottom of the dam is reduced, currents become uncontrollable and the horizontal thrust on the dam reaches a dangerous level. Erosion in rock-fill and earth-fill dams weakens them appreciably. The failure takes place commonly if the dams are made up of filled material. On 12 July 1961, a high flood caused the Panshet dam on the Mula river to topple. The surging water gushed down the valley and toppled the Khadakvasla dam. The turbulent waters from two successive reservoirs submerged the city of Pune within hours. The failure of the Machhu dam in Saurashtra, due to excessive rains in August 1979, killed more than 1,400 people and destroyed 12,000 houses in the Morvi township.

There is another phenomenon that has damaging effects on earthen embankments and dams—tunnel or piping erosion. Water seeps into the earthen structure through inter-granular pores and openings, such as those made by rats and other burrowing animals. The percolating water washes away sediments, eventually producing a network of tunnels. When these tunnels collapse, the structure is seriously damaged.

In order to ensure stability and safety of dams, it is imperative that very little or no water is allowed to seep under the foundation. Dams are, therefore, to be built on rocks that are free from structural discontinuities such as joints, fractures, faults and shear zones, and do not suffer from compositional infirmities such as porosity-permeability, or a cavernous nature that will let water seep underneath the dam. Since most of these infirmities are invariably present, the rocks need to be treated adequately to be strengthened. If there is no option, the dam should be aligned parallel to the strike of these features, as far as possible, so that only a small portion of the problematic foundation lies under the dam.

Dams are located in such a manner that the impervious layers or blocks form an apron on the edge of the dam (Figure 17.3). The impervious apron prevents seepage towards the base of the dam. This was done in the case of the Bhakra dam on the Satluj river in Himachal (Figure 17.6). The most favourable condition is offered by upstream-dipping, impervious beds (Figure 17.3), as is the case in the Kalagarh dam on the Ramganga river in Uttaranchal, and the Pong dam across the Beas river in Himachal. However, steeply downstream-dipping beds provide a comparatively better condition (Figure 17.3B) than gently downstream-dipping beds. To locate dams along the crestal plane, synclinal structures are less favoured than anticlinal ones (Figure 17.3C), because water percolating through the permeable beds may extend pressure on the dam. If the natural conditions are not favourable, water-tightness is achieved by placing an artificial apron or blanket of impervious material, like clay, cement-concrete, gunite, asphalt, etc on the upstream side of the foundation, or by constructing **cut-off** structures (Figure 17.1C and 17.2).

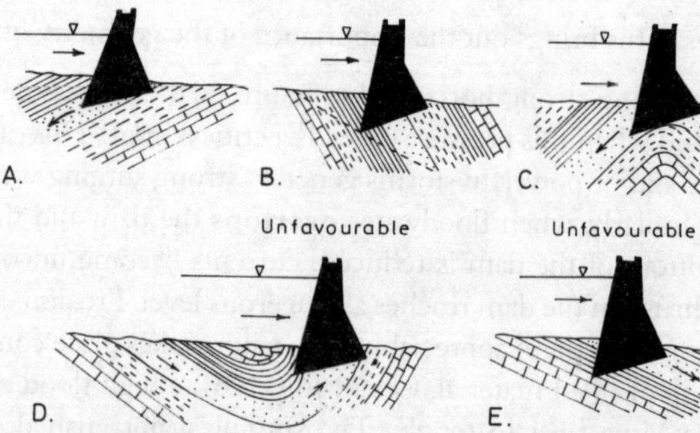


**Figure 17.3.** The location of a dam (solid block): (A) and (B) apron of impermeable shale or clay prevents water from seeping under the dam; (C) permeable layer carries the seeping water away from the site of the structure; (D) water percolating through the permeable layer comes under the dam and makes it vulnerable to buoyance-uplift; (E) dam loc: ted on a permeable layer is exposed to the action of seeping water. (The dam is shown in black, the lined beds are made of impervious rocks, while the dotted are permeable rocks).

The frictional resistance between the dam and its foundation is an important factor in its safety. The presence of water and clay imparts slipperiness to the foundation, thus undermining the stability of the dam. Therefore, weathered rocks characterized by clayey material are not suitable for erecting dams. Weathered rocks are invariably removed from the foundation of concrete and masonry dams, and for earthen dams, the weathered rocks are stripped of the clay-rich regolith wholly or made very rough on the surface. The structures are then suitably keyed to certain depths by constructing a pile foundation and **cut-off trenches** at the base of the core trenches (Figure 17.6D). This was done for the Kalagarh and Pong dams. In Karnataka, highly weathered Precambrian rocks—weathered upto a depth of 50 m or more—form the location of the 54-m Linganmakhi dam on the Sharavati river, in the central Sahyadri. Nearly 20 m of thick clay-rich weathered material was removed from the riverbed on the right bank (Figure 17.4). In order to overcome problems, a 33-m deep pile foundation was constructed before erecting the dam. The soil being permeable, it could have acted as a conduit for impounded water to escape. It was therefore removed before constructing the dam.

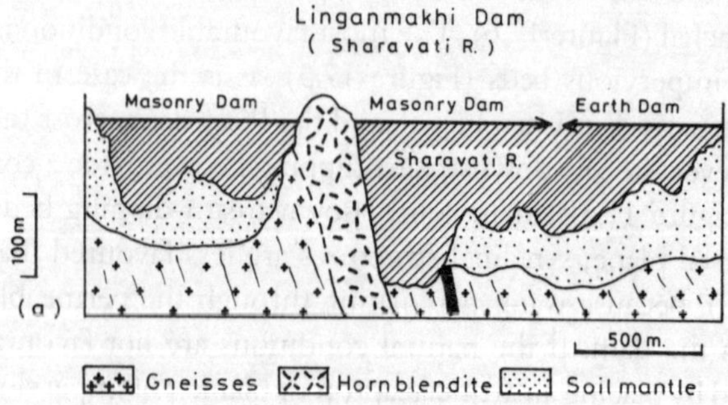


**Figure 17.4.** Before the 54-m Linganmakhi dam was constructed, nearly 20 m of thick clay-rich weathered rocks and soil were removed. The dam is keyed to the foundation by making a 33-m deep pile foundation (After Murthy and Srinivasa, 1972).

## TREACHEROUS FAULTS AND SHEAR ZONES

Faults are invariably associated with shear zones (Figure 17.5D and 17.6). These are characterized by splitting, shearing, and shattering or crushing of rocks. Understandably, the rocks of these zones have a low bearing capacity. Moreover, they provide easy access and passage to percolating water. Therefore, dams built on shear zones are exposed to threats of instability. Likewise, cavernous rocks such as limestone and dolomite make poor foundations and are best avoided.

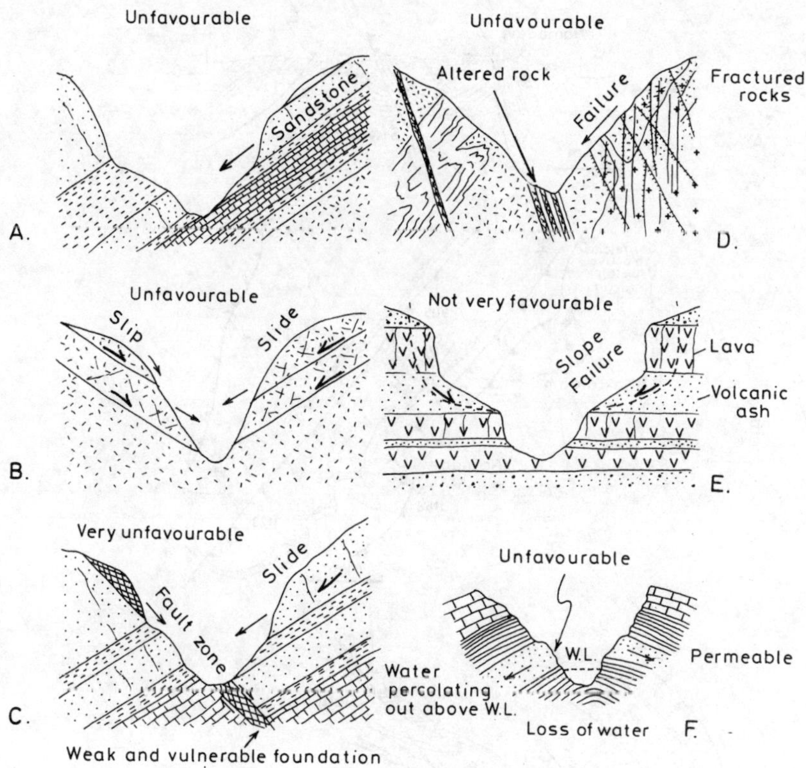


**Figure 17.5.** (A) (B) (C) Dams located in these valleys are vulnerable to landslides. The fault zone further adds to the weakening of the riverbed foundation; (D) The valley is prone to rock-falls and slips due to a multiplicity of fractures. The altered rocks make poor foundation; (E) Slope failures are common on valley slopes made up of jointed basalts, particularly where they rest upon incompetent volcanic ash beds; (F) If the dam is located along the crest of an anticline made up of permeable beds, the impounded water will easily seep out in large quantities.

Faults crossing a valley do not pose much risk unless they are active—that is, unless movements have taken place along the fault planes in the last few thousand years, or if they occur within the foundation of the dam. The 226-m Bhakra dam is located close to a thrust and a cross fault (Figure 17.6). The repetition of beds and the elimination of a clay bed are attributed to movement along the faults. In order to overcome the problem of shear-zone instability, weakened material upto a depth of 45 m was excavated and the shear zones **back-filled** with pre-cooled concrete.

This was done to overcome the problem of differential settlement. Incidentally, back-filling with concrete also serves as a key to prevent the dam from sliding. Extensive cement **grouting**—the forcible injection of fluidal cement by pressure—down to depths of 5–9 m strengthened the foundation appreciably. A 90-m deep **grout curtain** near the heal provided additional protection against the action of water (Ramachandran and Gangopadhyay, 1972).

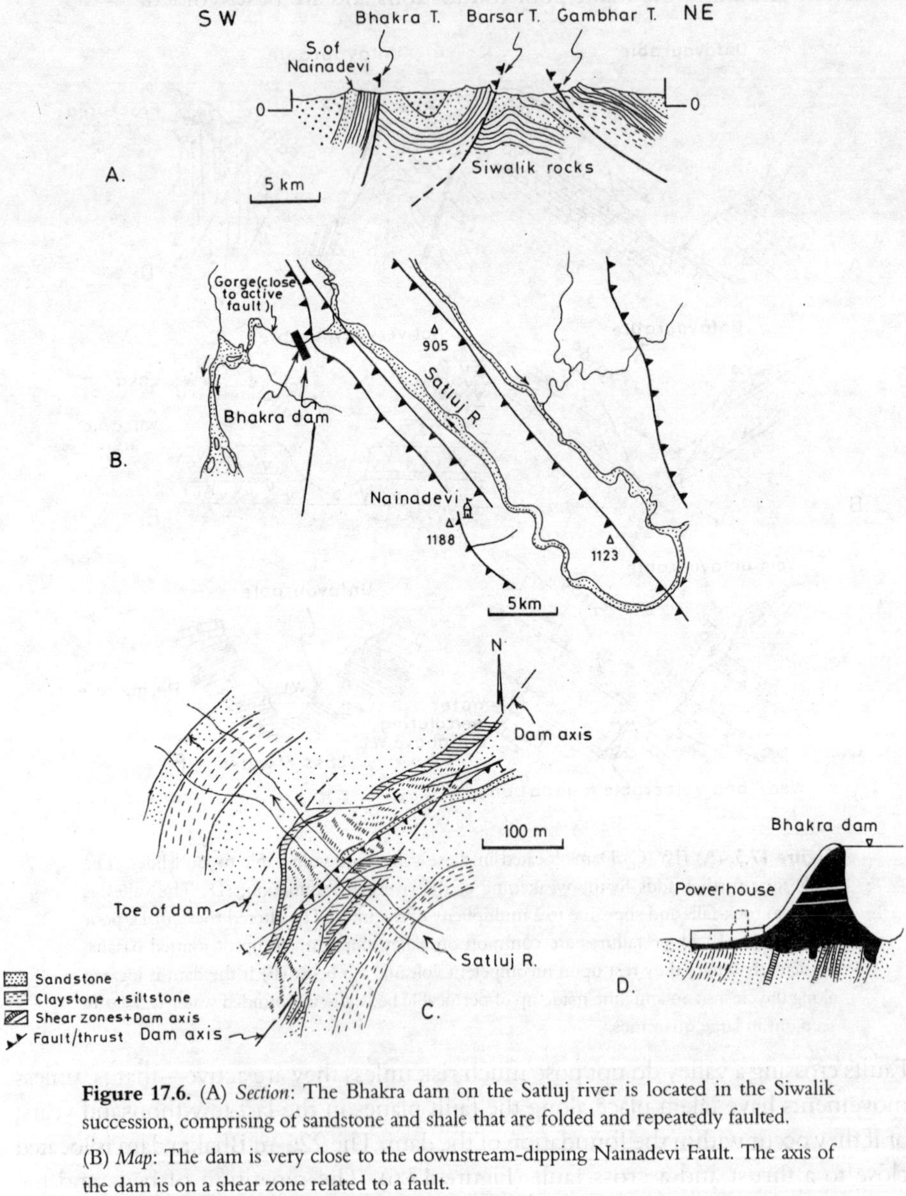


**Figure 17.6.** (A) *Section*: The Bhakra dam on the Satluj river is located in the Siwalik succession, comprising of sandstone and shale that are folded and repeatedly faulted.

(B) *Map*: The dam is very close to the downstream-dipping Nainadevi Fault. The axis of the dam is on a shear zone related to a fault.

(C) The shear zones and fault-gouge belts were excavated and back-filled with cement concrete.

(D) The dam is deeply keyed to the foundation, and the steeply-dipping impervious clay beds at the toe prevent water from getting underneath the dam.

In the Bhagirathi valley in Garhwal, the Chandpur phyllite succession is characterized by a multiplicity of 10–30 cm (and as much as 11–22-m wide) shear zones. It forms the foundation of the 260-m high Tehri dam (Nawani et al, 1991). Intensive grouting was carried out to seal and plug the zones of weakness (Figure 17.7). The Nagarjunsagar dam foundation is beset with a 5-m wide fault zone in the gneissic granite on the left abutment, and a 1.5-m wide shear zone across the dam axis. Excavation and back-filling with concrete was done before the dam was erected. The provision of an 8.5-m deep cut-off trench near the upstream toe, with intensive grouting upto a depth of 73 m, strengthened the dam's foundation in the Krishna riverbed.

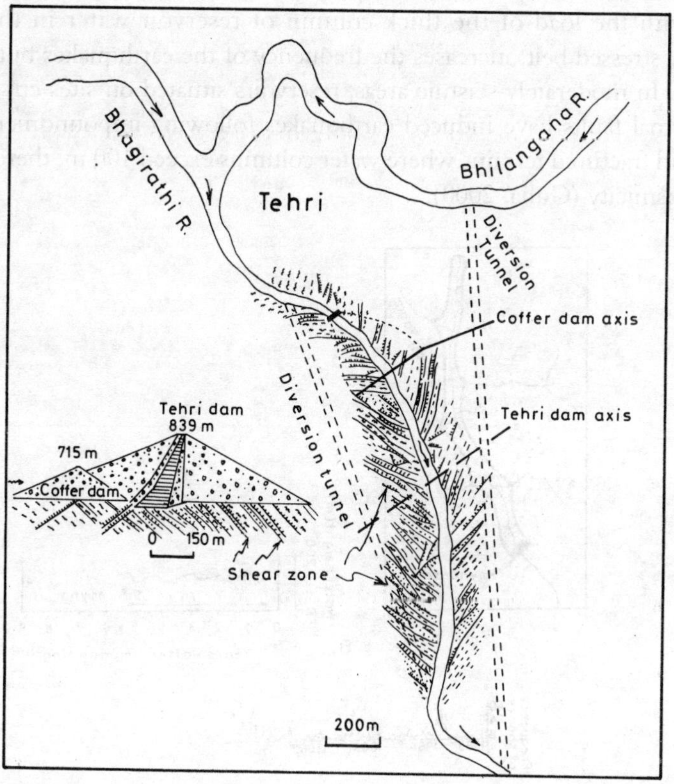


**Figure 17.7.** The Tehri dam on the Bhagirathi river is located on a foundation of phyllite alternating with quartzite. The foundation is characterized by a multiplicity of shear zones dipping upstream. To the left is the section of the 260-m high earth-fill dam with a clay core (Based on Pant et al, 1980).

Another problem of equal seriousness is created by cavernous limestone, dolomite or marble, particularly in areas of karst topography. The Obra dam on the Rihand river in U.P.; the 42-m Kamthikhere dam on the Pench river in Nagpur district (Maharashtra); and the 730-m Bajaj Sagar dam on the Mahi river in Gujarat faced serious foundation problems, requiring very special treatment (Sundaraiya and Wakpanjar, 1996). The remedial measures included constructing a concrete diaphragm (0.6-m thick, 3-m apart) across the entire thickness of the foundation rocks, entrenched deep into the underlying bed; and grouting the weakened rocks, building a grout curtain along the central line. The weakened cavernous rock was removed altogether and back-filled with concrete.

## ACTIVE FAULTS AND SEISMICITY

The 104-m high Koyna dam is located in a valley that follows a 2–4-m wide shear zone trending NNE–SSW. The shear zone in the terrain of basaltic lavas is related to seismogenic faults as borne out by frequent seismicity in the region (Figure 17.8). Part of the seismicity is what is described as **reservoir-induced seismicity** (Gupta, 1992; Guha, 2000; Talwani, 2000). The ingress of water from the impoundment, coupled with the load of the thick column of reservoir water in the already tectonically stressed belt, increases the frequency of the earthquakes but not their magnitude. In moderately seismic areas, reservoirs situated on sites cut by strike-slip or normal faults have induced earthquakes following impoundment; and in volcanic and fractured terrains where water columns exceed 100 m, there has been induced seismicity (Guha, 2000).

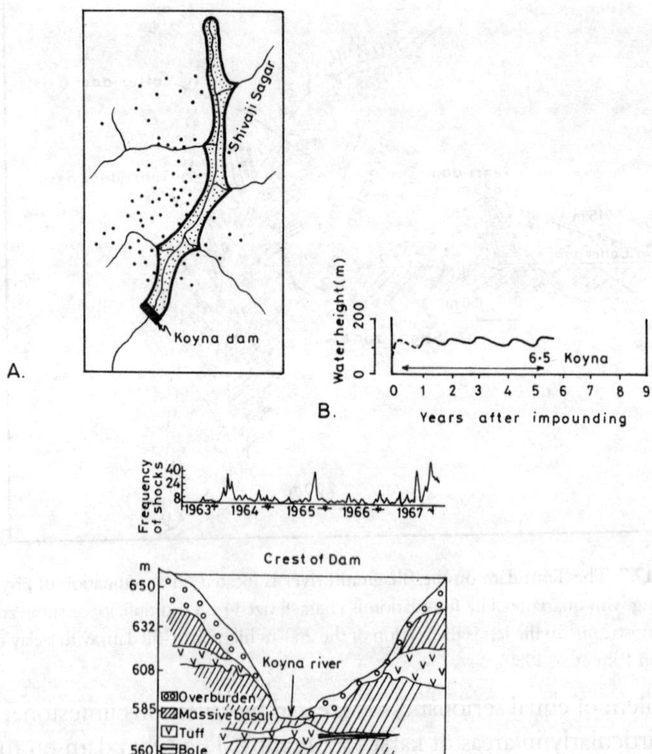


**Figure 17.8.** (A) Shivajisagar (reservoir) behind the Koyna dam induced seismicity as reflected in the distribution of epicentres (dots)

(B) Graph (upper) shows frequency of earthquakes following water impoundment.

(C) The (lower) section shows the lithology of the Koyna area.

It may be emphasized that the collective presence of three conditions is necessary for reservoir-induced seismicity:

(i) Excess load of water and its infiltration to the deeper critical part through fissures, fractures, faults or holes,

(ii) The prevalence of stress build-up in the interior where the *active* fault is locked, and

(iii) The existence of brittle rocks that will not yield, bend or flow, but break or snap.

This explains why there is no perceptible seismicity in the areas of the Bhakra, Mangla and Tarbela dams in the Outer Himalayan domain made up of sandstone alternating rhythmically with plastically yielding shale, mudstone and clay.

It may be recapitulated that active faults are episodically reactivated, resulting in earthquakes. The **ground shaking** that follows earthquakes counteracts the vertically downward-acting force due to the weight of the dam. This implies that the acceleration due to gravity (g) suffers *deceleration* as a consequence of the ground motion. This condition is bound to destabilize the dam in the earthquake-prone area, in the epicentre. If the peak ground acceleration exceeds the value of g, the dam could topple. It may be emphasized that the ground motion during an earthquake is not just in a vertical direction, countering the weight. Primarily the ground motion that is determined is the horizontal component (R.N. Iyengar, per. com., 2002). The Tehri dam in the Garhwal Himalaya has been designed to withstand the peak ground acceleration of 0.446 g that would be generated by an earthquake of magnitude M 7.2. In the event of an earthquake of magnitude of M≥8, the spectral acceleration in the field area could be as high as 1.0 g (Khattri, 1995).

Apart from ground acceleration, the passage of ground waves may cause alternate opening and closing of fissures and shear zones in the foundations of dams. This might allow ingress of water and thereby increase the pore-water pressure under them (Valdiya, 1997). Moreover, the earthquake may trigger landslides on a large scale, even threatening the dam and its reservoir. In snowy domains, avalanches are triggered by earthquakes.

The Narmada river follows a rift valley—an elongated depression formed due to sinking of the ground between two nearly parallel faults. The seismicity of the belt (Chapter 8) indicates that the boundary faults are quite active. This is borne out by recent landforms that were offset and displaced due to movements on the active faults. A multiplicity of dams—including the 129-m high and 1.2-km long, concrete gravity Sardar Sarovar dam at Vadgam (with a storage of 5800 million $m^3$), located in a lava terrain, which is cut by ENE–WNW oriented dolerite dykes and a steep-angled 7-m wide zone of reverse fault (Pancholi, 1997)—are being built across the Narmada in Gujarat and Madhya Pradesh (Figure 17.9). It remains to be seen how these dams will behave when larger earthquakes, or more frequent moderate earthquakes, occur in the Narmada rift valley.

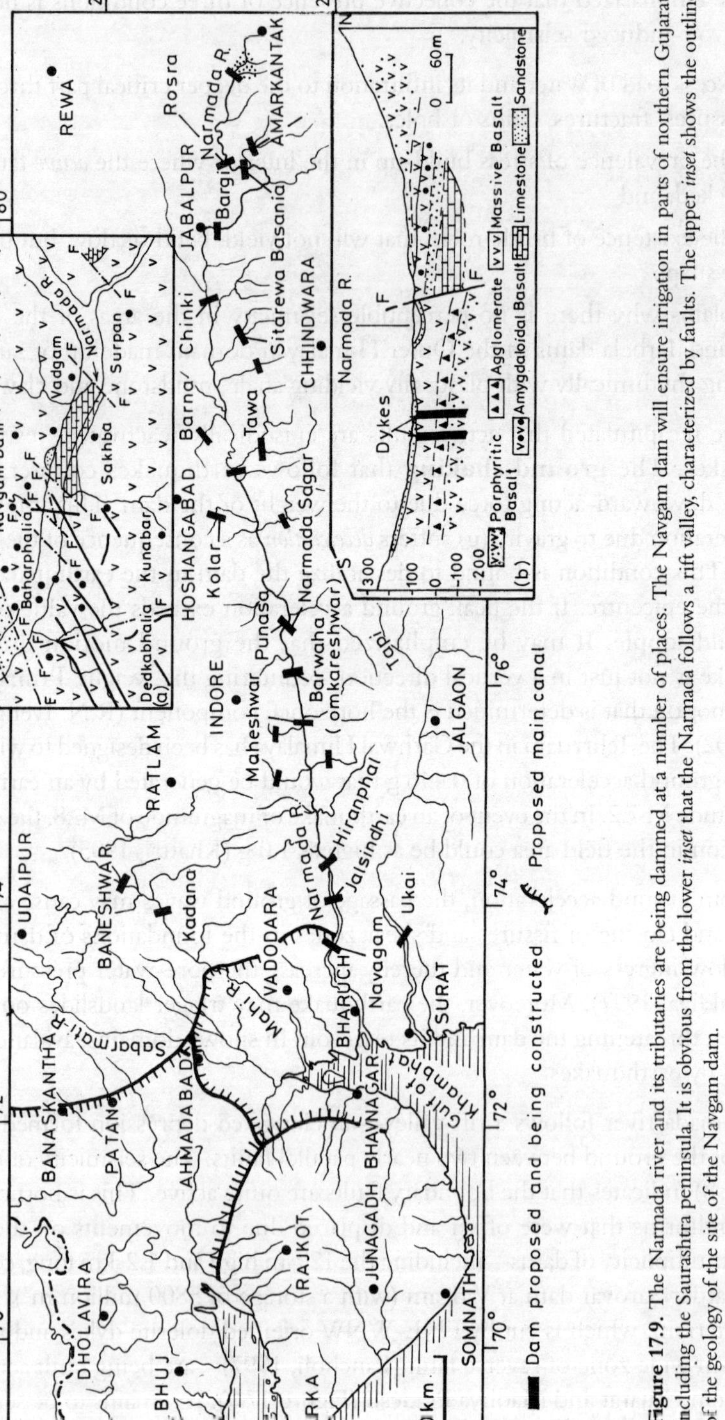


**Figure 17.9.** The Narmada river and its tributaries are being dammed at a number of places. The Navgam dam will ensure irrigation in parts of northern Gujarat, including the Saurashtra peninsula. It is obvious from the lower *inset* that the Narmada follows a rift valley, characterized by faults. The upper *inset* shows the outline of the geology of the site of the Navgam dam.

## Ecological Impact of River Impoundment

River impoundment causes stagnation of a large body of water, and this in turn entails evaporation, and therefore, increased vapour content in the air above the reservoir and its surrounding regions. It is estimated that nearly 12.3 million m$^3$—equivalent to 1.2-m columns—of water in the Govindsagar behind the Bhakra dam is annually lost due to evaporation. The increase in the content of vapour in the air causes microclimatic changes within the limited extent of the river basins. The summers become shorter and more humid and the winters less severe. The net loss of heat available to plants, appreciably affects their growth. The larger the surface of the water body, the greater is the moderation of the microclimate.

The change from a flowing-water regime to stagnant-water ecosystems manifests itself in a change in the quality of water and the lives of the organisms, both aquatic plants and animals. Organic matter washed down from the vegetated slopes, and derived from the submerged forests, pastures and agricultural fields, not only pollutes the reservoir water, but also promotes vigorous growth of certain organisms, including those that harbour disease-carrying organisms. In Raichur district of Karnataka, for example, the occurence of malaria has spread since the dam across the Tungabhadra river was built. The superabundance of organic matter in the reservoir water promotes vigorous microbial activity, leading to the explosive growth of some organisms in the initial stage of impoundment. The prolific growth of weeds like hyacinth is a case in point. In the Govindsagar, while fish like *gid* became prolific, others like *gugli*, *mahseer* and *kunni* dwindled or disappeared.

Many reservoirs in Andhra Pradesh and Tamilnadu, including the Nagarjunsagar, impound water that is enriched with fluorine, derived from the fluorite-bearing pegmatites in the lake basins. The use of these waters is responsible to a great extent for a very high incidence of *fluorosis,* a crippling bone-disease that affects a sizeable section of the region's underprivileged people, whose staple diet includes crops irrigated by the fluorine-enriched reservoir water (see chapter 19).

### SLOPE DESTABILIZATION ALONG RESERVOIR RIMS

The rise of water level in reservoirs causes the water table in the surrounding areas to move up. As groundwater seeps into the slopes of the basin, there is reduction in the shear strength of the rocks/soils that form the slopes. The drawdown, on the other hand, lowers the water table and alters the hydrological condition underground. Moreover, lowering of water level in reservoirs means withdrawal of the lateral support that the reservoir water gives to the slopes. Frequent rise and fall of the water level (resulting from frequent filling and emptying of the reservoir) creates stress fluctuation in the slope material. This causes instability and attendant failure of the slopes (Figure 9.4). It must be borne in mind that slope instability is invariably governed by a few factors like: (i) quality of slope-forming material, (ii) relief of the slope, and (iii) geometry of the material of the slope. If unfavourable factors prevail, the hill slope boundary of reservoirs will be affected by landslides. The debris that reaches the reservoir forms colluvial fans and deltas within the reservoir.

## SEDIMENTATION IN RESERVOIRS AND THEIR SHORTENED LIFE

The sediment load of rivers and their tributary streams is deposited in the still waters of the reservoir, forming deltas where the streams meet the reservoir. The deltas thus formed grow progressively downstream (Plate 20), eventually overwhelming the whole of the reservoir. The Kulekhani reservoir, behind a 150-m high dam in southcentral Nepal, is progressively losing its water-holding capacity due to its rapid rate of sediment accumulation (Figure 17.10). Its original storage capacity of 85.3 million m$^3$ in 1982 dwindled to 78.3 million m$^3$ by July 1993, due to sediment that was being eroded from the catchment at a rate of 1,700 m$^3$/km$^2$/yr. During a single rainstorm on 19 July 1993, about 4.8 million m$^3$ of sediment was deposited, the rate of erosion in the catchment being 31,700 m$^3$/km$^2$/yr (Galay et al, 1995).

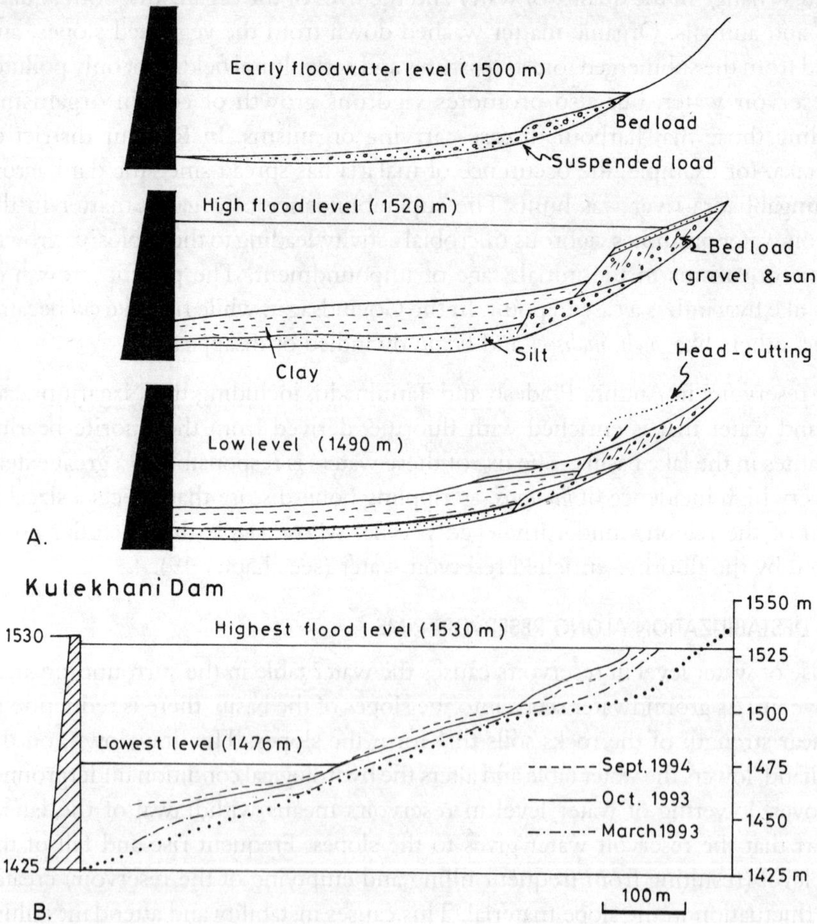


**Figure 17.10.** (A) Trend of deposition of sediment in the Kulekhani reservoir, SW of Kathmandu in southcentral Nepal

(B) Profile of the riverbed in the Kulekhani reservoir, illustrating accumulation of sediments in three successive periods (After Galay et al, 1995).

As a consequence of the quickened pace of erosion in catchment areas and its resultant larger input of sedimentary debris (at rates 1.5–3 times more than the rate anticipated at the designing stage), the water-holding capacity of reservoirs is consequently decreasing progressively—at rates of 0.30–0.90% annually (Lagwankar et al, 1994). The all-India siltation rate varies between 0.15 m ha/100 km²/yr and 27.85 m ha/100 km²/yr (Lagwankar et al, 1994). An estimate indicates that the economically useful lifespans of most artificial reservoirs has been reduced to half or a quarter of what was originally calculated by the planners of the dams (Table 17.1). The Pandoh reservoir in the Beas river lost 2.4 million m³ of its total gross storage capacity of 41 million m³, in the period 1977–96 (Nanda et al, 1998). The storage capacity of the Neyyar reservoir in Kerala declined by 28.8 million m³ in 38 years, at an annual rate of 0.71% (Suresh et al, 2000).

**Table 17.1.** Sediment accumulation in some reservoirs (After Lagwankar et al, 1994).

| Reservoir (River) | Year of Impoundment | Rate of Sedimentation (ha-m/100 km²/yr) | | | Percentage of Annual Loss of Storage Capacity |
|---|---|---|---|---|---|
| | | Assumed | Observed | Year | Gross |
| Pong (Beas) | 1974 | 4.29 | 27.85 | 1986 | 0.41 |
| Bhakra (Satluj) | 1958 | 4.29 | 5.66 | 1987 | 0.32 |
| Kalagarh (Ramganga) | 1974 | 4.25 | 22.94 | 1986 | 0.33 |
| Mayurakshi (Mayurakshi) | 1955 | 3.75 | 6.83 | 1975 | 0.51 |
| Matatila (Betwa) | 1956 | 1.33 | 5.45 | 1990 | 0.71 |
| Gandhisagar (Chambal) | 1960 | 3.57 | 8.96 | 1976 | 0.29 |
| Ukai (Tapi) | 1974 | 1.49 | 7.16 | 1984 | 0.53 |
| Koyna (Krishna) | 1961 | 5.05 | 8.10 | 1986 | 0.025 |
| Nizamsagar (Marjina) | 1930 | 2.38 | 4.89 | 1975 | 1.41 |
| Tungabhadra (Tungabhadra) | 1953 | 4.29 | 6.48 | 1985 | 0.49 |
| Panchet (Damodar) | 1956 | 6.67 | 5.88 | 1985 | 0.4 |
| Maithon (Barakar) | 1956 | 1.62 | 13.02 | 1971 | 0.5 |
| Hirakud (Mahanadi) | 1957 | 2.50 | 6.62 | 1984 | 0.61 |

The impoundment of water causes, among other things, entrapment of sediments so that the water that flows out of a reservoir is devoid of bed-load sediments and suspended sedimentary particles. The water discharged from the reservoir, therefore, has stronger propensity to erode deeper into the channel beds. In other words, downstream of dams, the channels show accelerated erosion. This gives rise to gully formation or deepening of the channels. Due to damming in the upper reaches, sediment-starved rivers in Kerala do not have enough sedimentary load to dump at their mouths and are unable to contribute to the building of beaches. (The sediment deposited at river mouths is a natural provider of material for beaches.) In the absence of beach nourishment by rivers, the Kerala coast has become increasingly vulnerable to coastal erosion.

## LEAKAGE AND WATERLOGGING

The stored water in reservoirs infiltrates down into the ground and raises the water table in the area surrounding the water bodies. Not only is there a loss of water through leakage, there is also the problem of waterlogging in low-lying terrains. The leakage and extent of waterlogging depends on the permeability of rocks and/or soil, or the embankment material. The geological setting permitting, waterlogging may lead to the development of marshy conditions. This harmful fallout of water impoundment is witnessed (Table 17.2) in the low-lying areas associated with the Hirakud dam in the Mahanadi floodplain in Orissa, the Raṇa Pratap dam and Kota dam in the Chambal plain in Rajasthan, and the Umrong dam in the Kopili floodplain in Meghalaya, to give a few examples.

**Table 17.2.** Problem of waterlogging and salinization associated with water impoundment (After Indian Institute of Public Administration, 1988)

| Reservoir | Areas Affected (in ha) | |
|---|---|---|
| | Waterlogged | Salinized |
| Jaisamand | — | 606 |
| Bhakra | — | 171,000 |
| Paithan | 215 | 660 |
| Vani Vilas Sagar | 276 | — |
| Siddheshwar | 558 | 4990 |
| Malaprabha | 1,146 | 139 |
| Hidkal | 3,521 | 289 |
| Yeldari | 9,595 | 316 |
| Tawa | 14,000 | — |
| Rana Pratap Sagar, Stage I | 16,402 | — |
| Sriramsagar | 20,000 | — |
| Tungabhadra | 41,000 | 81,000 |
| Nagarjunsagar | 57,000 | 23,000 |
| Gandhisagar | 70,000 | 64,838 |
| Total | 233,713 | 346,838 |

## SUBMERGENCE OF NATURAL RESOURCES

The most precious resources commonly lost due to submergence under reservoir waters are forests and agricultural fields in the basiṇ upstream of the dams (Figure 17.11). In many cases, valuable mineral deposits were also lost. The higher the dam, the greater the loss. Table 17.3 shows the extent of forests lost—nearly 401,400 ha until about the mid-seventies (Sharma, 1977).

**Table 17.3.** Forest lost during a short period of 22 years (1951–73) on account of water resources development (Sharma, 1977).

| State | Forest lost (ha) |
|---|---|
| Andhra Pradesh | 27,900 |
| Assam | 19,700 |
| Bihar and Jharkhand | 1,300 |
| Gujarat | 34,700 |
| Himachal Pradesh | 7,600 |
| Jammu & Kashmir | 100 |
| Karnataka | 28,700 |
| Kerala | 7,800 |
| Madhya Pradesh and Chhattisgarh | 69,000 |
| Maharashtra | 10,300 |
| Orissa | 32,500 |
| Rajasthan | 14,500 |
| Tamilnadu | 45,100 |
| Tripura | 7,400 |
| Uttar Pradesh and Uttaranchal | 93,200 |
| Arunachal Pradesh | 100 |
| West Bengal | 1,300 |
| All India Total | 401,400 ha |

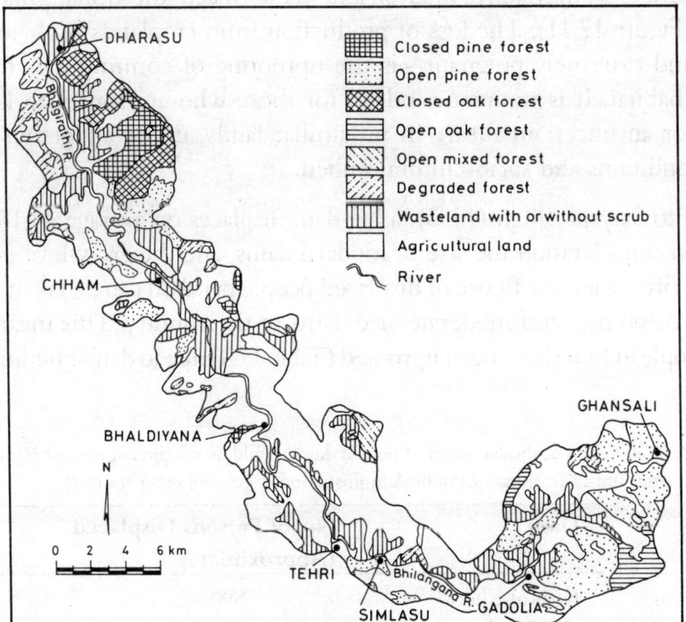


**Figure 17.11.** Loss of natural and artificial resources due to anticipated submergence when the Tehri dam on the Bhagirathi river rises to its full height of 260 m (After Garg et al, 1995).

Apart from this submergence of forests, clear-felling of forests on a large scale—to build residential colonies that house project personnel, construct road networks, and supply fuel wood for the labour forces—has caused extensive damage to forests throughout India.

It must be stressed that natural resources are also capital resources, with economic value. While calculating the costs for projects that develop water resources, all natural resources must be costed. Stored water is a capital resource, and must be appropriately costed. The price at which this water is supplied for agriculture is highly subsidised. This is not sound economics.

These resources also include forests, soils, freshwater bodies, and horticultural and agricultural farms. This understanding has to be incorporated in politico-economic planning and decision making. Before embarking on any project, it is essential to ask if our children will inherit as much as we did from our parents. In the endeavour for development, the obligation to future generations must not be forgotten.

## UPROOTING PEOPLE: SOCIO-CULTURAL DISTRESS

Fluvial terraces in mountains, and floodways in the plains, are primary regions for cultivation. The spread of reservoir water brings about total submergence of these agricultural tracts, and the shelter it gives to rural communities. The problem becomes extremely serious where the population density is high, as in the Narmada valley (Figure 17.9) in Madhya Pradesh and in the Bhagirathi–Bhilangana valleys in Garhwal (Figure 17.11). The loss of production from croplands is obvious. More critical—and extremely poignant—is the uprooting of communities from their preferred habitat. It is traumatic indeed for those who are forced to leave their homes, for an uncertain future in unfamiliar lands, under altogether different climatic conditions and socio-cultural milieu.

According to Satyajit Singh (1993), a big dam displaces on average 44,182 people. Taking into consideration the size of modern dams, and the growth of population in the past fifty years, the figure of displaced people per dam can be taken at 10,000. There are 3,300 big- and moderate-size dams in the country. This means that 33 million people in India have been uprooted (Table 17.4) due to dam-building projects (Roy, 1999).

**Table 17.4.** Extent of displacement of people due to building of dams in the last 50 years (Data from Indian Institute of Public Administration, 1988; and other sources)

| Dam | No. of Persons Displaced (Approximate) |
|---|---|
| Kamthi Kheri | 890 |
| Ozarkhad | 1,350 |
| Bagh | 1,370 |
| Karanjwan | 1,600 |
| Palkhed | 1,720 |
| Ibadeh | 2,260 |

| Dam | No. of Persons Displaced (Approximate) |
|---|---|
| Bennithora | 3,000 |
| Kenher | 7,080 |
| Kayadhu | 8,860 |
| Kabibi | 11,250 |
| Hemavathy | 11,600 |
| Nagarjunsagar | 13,230 |
| Krishnarajasagar | 15,000 |
| Isapur | 15,590 |
| Dhimbe | 18,000 |
| Hirakud | 18,000 |
| Dham | 19,730 |
| Almatti | 20,000 |
| Mahi Bajaj Sagar | 27,320 |
| Warna | 29,300 |
| Koyna | 30,000 |
| Gohira | 31,000 |
| Hidkal | 31,130 |
| Bhima | 35,070 |
| Bhakra | 36,000 |
| Tehri | 86,000* |
| Narayanpur | 48,120 |
| Gandhisagar | 51,510 |
| Ukai | 52,000 |
| Srisailam | 52,050 |
| Rihand | 55,000 |
| Balimola | 60,000 |
| Majal Gaon | 65,300 |
| Nizamsagar | 67,440 |
| Sriramsagar | 75,090 |
| Ukai | 80,000 |
| Tilaya | 13,450 |
| Maithon | 28,030 |
| Panchet | 41,460 |
| Tungabhadra | 54,450 |
| Kangsabati Kumari | 125,000 |
| Pong | 16,000 |
| Subanrekha | 120,000* |
| Koel | 98,000* |
| Sardar Sarovar | 400,000* |
| Navagam | 150,000* |
| Maheswar | 400,000* |
| Bargi | 70,000* |

* Expected if or when the dams are built.

## Smaller Dams: A Better Alternative

As bigger dams engender extensive environmental imbalance and cause societal distress, including uprooting of people (Valdiya, 1992; 1997) and since there is, as of now, no built-in provision for mitigation or elimination of negative environmental and socio-cultural effects, it would be better to invest in smaller dams. Smaller dams cost less, start giving benefits quite early, bring greater profits, and cause less damage to or impairment of environment, croplands and habitats. This is quite evident from the data (Table 17.5) collected by the Planning Commission (Singh, 1993).

**Table 17.5.** Expenditure on and benefits from major and minor irrigation projects: Planning Commission Statistics (Singh, 1993).

| National Plan | Quality/Expenditure (million rupees) | | Cumulative Potential (million hectares) | |
|---|---|---|---|---|
| | **Major/Medium** | **Minor** | **Major/Medium** | **Minor** |
| Pre-plan benefits | — | — | 9.7 | 12.9 |
| First Plan | 3,800 | 760 | 12.2 | 14.06 |
| Second Plan | 3,800 | 1,420 | 14.3 | 14.79 |
| Third Plan | 5,810 | 3,280 | 16.6 | 17.01 |
| Annual Plan 1966–1969 | 4,340 | 3,260 | 18.1 | 19 |
| Fourth Plan 1969–1974 | 12,370 | 5,130 | 20.7 | 23.5 |
| Fifth Plan 1974–1978 | 24,420 | 6,310 | 24.82 | 27.3 |
| 1978–1979 | 9,770 | 2,370 | 25.86 | 28.60 |
| Annual Plan 1979–1980 | 10,790 | 2,600 | 26.6 | 30 |
| Total | 75,100 | 12,930 | 168.88 | 187.16 |

Admittedly, the cost of electricity production is high (nearly double) for small hydel projects, as compared to that of large projects. The cost of water storage is equally high. However, if one were to think of the real cost of water resources development—which includes the value of the lost natural resources, the cost of restoration of damaged land and the expenditure on the relocation/rehabilitation of uprooted communities—this high cost of electricity generated by small hydel projects is more than compensated (Valdiya, 1992; 1997). An example of an environment-friendly modern dam is the 28-m high and 95-m long structure in the Jhelum river near Buniyar, in Baramula district in Kashmir. This run-of-the river project, with a large component of underground works, will have the capacity to generate 480 MW of electricity. Located on the Satluj river in Himachal Pradesh, the 67-m high Nathpa dam of the Nathpa–Jhakri Project was responsible for uprooting only 52 families (Mande and Vohra, 1988).

India has an old (some dating back to the second and third centuries AD) and very successful tradition of irrigation, with water held in tanks developed in streams and rivulets, even in rain-deficient regions. Nearly 3.25 million hectares of land were irrigated from waters impounded behind small dams, before the advent of British rule and the Western concept of high dams.

There were 37,000 tanks covering 40 km² in Mysore State; 53,000 tanks irrigating 1.4 million ha in Madras Province; and 50,000 tanks irrigating 0.26 m ha in Madhya Pradesh (Kulkarni, 1982). The small dams that held water in these tanks showed 'a great boldness of design, [and] considerable knowledge of hydraulic principles', as one British engineer, R.B. Smith, wrote in 1856.

The Anangpur dam, built a thousand years ago, two kilometres south of Surajkund near Delhi, is a model of an environment-friendly dam that has benefitted the people of the Aravalli basin for over one millenium. The 12-m high, 20-m wide and 103-m long gravity dam, constructed of rocks—with a clear understanding of the inter-relationship between the various elements of nature (air, water, earth); and the relationship an engineering structure has to have with the local and wider environment, taking care to alleviate the problem of uplift pressure, silting and wave action—has withstood the ravages of earthquakes (Garg et al, 1999).

The Chinese model of tapping stream water is not only environmentally sound, but also economically rewarding. Small hydro-power projects entail cascade development (Figure 17.12) and multi-purpose utilization of water in streams and small rivers. These can be locally planned and developed by the people themselves. The investment required is less, the period of construction is short and the benefits come quite early. The people can fully utilize the stream drop, create upper reservoirs to increase the firm capacity of downstream power-producing stations, and operate the cascade stations jointly to maintain the required water balance. In such projects, the dams are not more than 30–50 m in height and each of them has the capacity of generating 6–12 MW of power (Naibo, 1986).

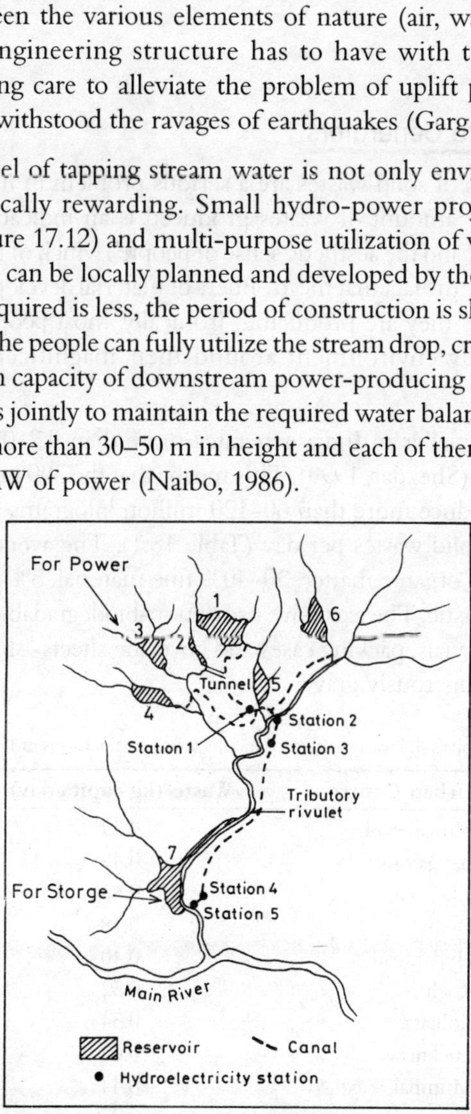


**Figure 17.12.** Tapping a small river in a mountainous terrain in China to generate electricity and store water for irrigation and flood moderation, using the cascade scheme (After Naibo, 1986).

# 18

# Disposing of Solid Wastes Without Polluting Groundwater

## Prolific Waste Generators

The mounting heaps of solid wastes are a serious problem in urban India. An old belief states that 'the amount of wastes produced is an indication of the level of cultural development and the aesthetic sense of people'. Much of urban India ignores this, as is borne out by the fact that the frontiers of material development are growing, but so are the wastes they are producing. Ironically, most people do not seem to mind this unfriendly environment around their magnificent new visions of architecture.

On average, a city-dweller in India generates more than 0.2–0.5 kg of municipal solid waste everyday (Shekdar, 1999). This means that the 300 million people living in urban centres produce more than 60–120 million kilograms (annually about 40 million tonnes) of solid wastes per day (Table 18.1). The average composition of this waste is 30–40% organic matter, 30–40% fine material, 5% paper and 1% each of metal, glass and plastic. The excessive use of non-biodegradable plastic material—such as bags, bottles, vials, packing cases and covering sheets—has made the waste-disposal problem dangerously grave.

**Table 18.1.** Per capita solid waste generated in selected municipal areas in 1994 (Vyas, 2001).

| Urban Centre | Waste (kg/capita/day) |
|---|---|
| Ahmadabad | 0.59 |
| Bangalore | 0.48 |
| Chennai | 0.66 |
| Delhi | 0.48 |
| Hyderabad | 0.40 |
| Kochi | 0.52 |
| Kolkata | 0.34 |
| Lucknow | 0.62 |
| Mumbai | 0.44 |
| Nagpur | 0.27 |
| Pune | 0.31 |
| Varanasi | 0.40 |

There are also enormous quantities of wastes produced by industrial units. For example, power plants produce flyash at the rate of 65 million tonnes per year (Sahu, 1998). Characterized commonly by toxic elements (such as nickle, chromium, mercury, cadmium, etc), voluminous mining wastes smother springs, choke channels and bury vegetation in areas of mineral exploitation.

In the rural world, solid waste is largely agricultural. Part of this waste is used as organic manure, part of it is incinerated and quite a large part is left *in situ* to decompose. In addition, human faeces accumulates at the rate of 51,000–71,500 tonnes per day, that is 18.6–26.1 million tonnes annually (Deodhar and Nath, 1981), much of which is left exposed to the air.

## Disposal of Wastes

### DEVELOPING LANDFILLS

Solid wastes must not remain exposed to the air and rain for long, nor be allowed to enter rivers. The only option is to bury the waste under impervious cover. This can be done by constructing scientifically sound landfills in geologically favourable sites. The waste is systematically filled into carefully chosen sites, covered each day with layers of earth, and then compacted. When the landfill is finally ready, an impervious cover of clay, asphalt or plastic is laid over the top to prevent the entry of rain or surface water. This is done in order to control the formation of **leachate** from soluble matter in the waste. Leachate can move far and wide if the floor—the geological formation—or soil below the landfill is permeable (Figure 18.1). It could percolate downwards and pollute the groundwater and the waters of springs and streams (Figure 18.2).

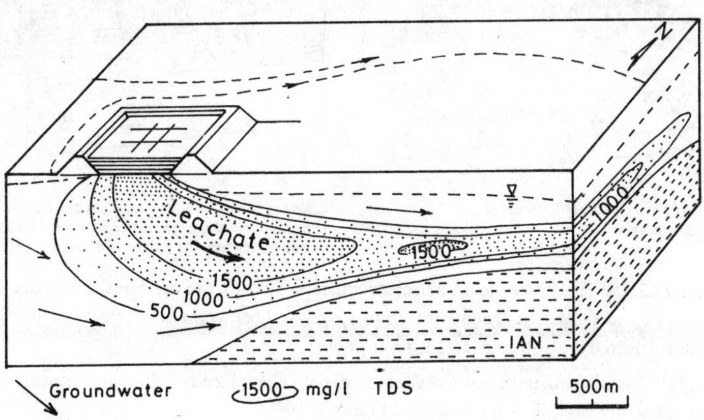


**Figure 18.1.** Once it gets inside a landfill, water dissolves soluble material in the waste and produces leachate. The leachate spreads far and wide in a base that is permeable. The contour lines show the amount of total dissolved substance (TDS) in milligrams per litre.

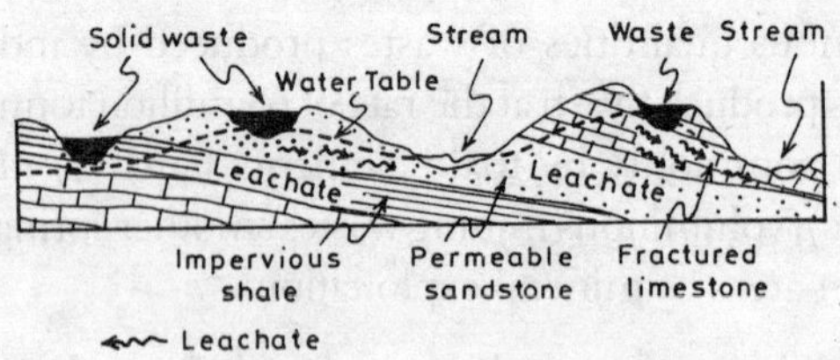


**Figure 18.2.** Landfills developed on permeable beds/horizon cause pollution of stream and spring waters, as the leachates find their way to them readily.

The following conditions must be met to develop an ideal landfill:

(i) The base (floor) of the landfill should be situated on an impervious material—rock or soil (Figure 18.3B). If this is not feasible, the base material should be treated properly to prevent or minimize percolation of water and the resultant leachate.

(ii) The landfill area should be free, as far as possible, from structural discontinuities such as faults, fractures, fissures, joints or open bedding that allow passage or flow of liquids.

(iii) The water table must be well below the base of the landfill (Figure 18.3A, 18.3B) so that the leachate, if formed, is not able to percolate down to the groundwater. Obviously, landfills must not be built in waterlogged tracts. This further implies that before a landfill is developed, the water table in the area should be lowered by intensive and continual pumping out of groundwater or by cutting deep trenches around the landfill.

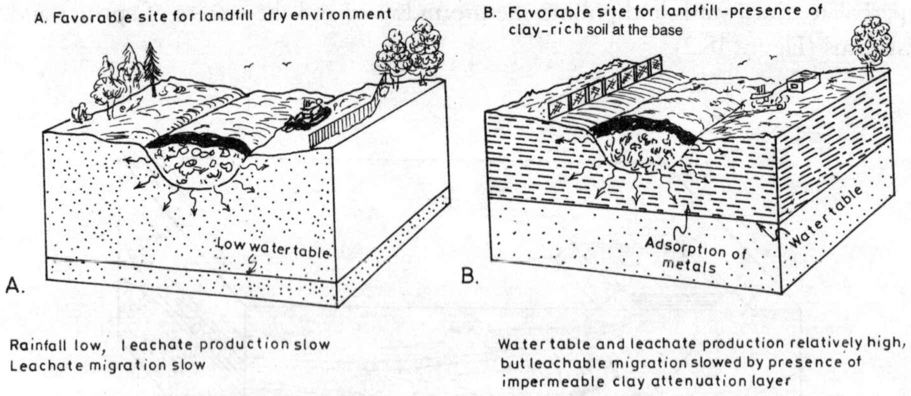


**Figure 18.3.** (A) Landfill made on permeable material. In order to prevent the formation of leachate, the water table has been lowered to a level well below the base of the landfill. The landfill is covered with an impervious layer of clay.

(B) Landfill is constructed on impervious material. The water table lies at a considerable depth below the landfill (After Marsh and Grossa, 1996).

## BIOLOGICAL PROCESSING OF SOLID WASTE

In rural areas, it is agricultural wastes and cattle droppings that form the dominant waste product. These agricultural wastes are either burned or utilized profitably. Organic waste, put in landfills or in pits, is commonly composted to create useful manure. The Tata Energy Research Institute at New Delhi has recently developed a process that entails acidification and methanation of agricultural wastes. Biomethanization of waste is done by the anaeorobic digestion of wastes. As the name implies, methane is produced in the process, which is useful for burning and power generation. The compost formed is a valuable manure.

## TACKLING NON-BIODEGRADABLE WASTE

Plastic material like packaging polythene, polyvinyl chloride pipes and sheets, plastic wares, etc constitute the most pervasive and repulsive solid waste. Since plastic material does not degrade biologically and its burning generates deadly dioxin gases, it is becoming extremely difficult to get rid of this waste. Burying is the only way. At present India consumes 4 million tonnes of plastic material annually, of which nearly 1.8 million tonnes become waste (Jamwal, 2002). Frighteningly, this amount could double in another five years.

## LAND DISPOSAL OF CHEMICALLY HAZARDOUS WASTES

The land disposal of chemically hazardous waste requires extreme care (Figure 18.4). The waste material is put in landfills that are provided with multiple layers of lining, which include clay and synthetic liners. The various types of synthetic liners include polyethylene, butyl rubber, neoprene and polyvinyl chloride (Shekdar, 1999).

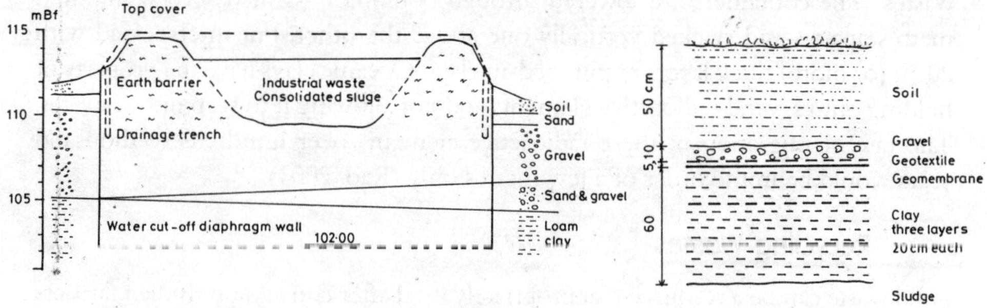


**Figure 18.4.** A design of the pit (trench) with multiple layers of lining and cover in which chemically-hazardous waste can be disposed of without causing any problem. The column to the right shows details of the lithology under the landfill. (After Pearce, 1983).

## DISPOSAL OF RADIOACTIVE WASTES

Radioactive solid and liquid wastes generated in India are growing in volume (Table 18.2). An estimate shows that, in India, while low-level nuclear wastes amount to over 232,000 m$^3$, intermediate-level and high-level wastes measure about 40,280 m$^3$—a total of more than 273,000 m$^3$ (Ramana et al, 2001). Low- and intermediate-level radioactive wastes are converted into slurry and then stored in concrete-lined containers. These containers are embedded in shallow land repositories that have concrete walls. In the case of high-level radioactive material, it is first vitrified in borosilicate glass or synthetic rock, and then stored in concrete-encased steel tanks (3 m deep). The containers can be kept in underground stores for twenty years before being disposed of in underground salt beds or in abandoned salt mines. These highly active wastes are also grouted in cement or bitumen and are carried to the disposal site in lead-shielded containers. High Integrity Containers are provided at the disposal site for additional containment of waste (Shekdar, 1999). Despite all these precautions, it must be admitted, there are still chances of leakage of gases and liquids.

**Table 18.2.** Average radioactive wastes annually generated in India (Rao, 2001)

| Source | Solid ($m^3$/yr) | Liquid ($m^3$/yr) |
|---|---|---|
| Research reactors | 23 | 16,000 |
| Power reactors | 180 | 53,600 |
| Fuel-processing facilities | 130 | 34,300 |
| R & D laboratories | 50 | 12,000 |
| Total | 383 | 115,900 |

Ocean trenches provide the most suitable sites for getting rid of high-level radioactive wastes. Since the ocean floor is subducting or sinking into the mantle (along with the material of the oceanic trenches), at a rate above 2.1 cm/yr, radioactive wastes would be swallowed by the mantle at a rate much faster than the rate at which the radionuclide elements diffuse (1 mm/yr) through the deep oceanic sediments (Rao, 2001). Parts of the ocean floor, which are deeper than five kilometres and blanketed with thick piles of clays, are safe places for burying containers containing radioactive wastes. The containers are lowered through boreholes—drilled several hundred metres apart—and stacked vertically one above the other, but interspersed with 20-m (or more) thick layers of pumped-in clays. Oceanic clays have the property of holding onto several radioactive elements and not allowing it to seep into seawater. The rate of migration of these radioactive elements over hundreds of thousand years would be in the order of a few metres only (Rao, 2001).

## Recycling Wastes

Solid waste can be a resource if appropriately used after conversion. Indian farmers have, from time immemorial, been converting cattle droppings into valuable manure, particularly by composting them. The solid waste generated by the urban population is recycled likewise in a variety of ways. An estimation shows that annually 1,429,281 tonnes of waste are recycled, and another 118,941 tonnes are amenable to recycling (Vyas, 2001). The municipal wastes in Mumbai are beneficially converted into briquettes and pellets, usable as a fuel of high calorific value. Human and animal excreta is used not only for the generation of biogas, but also to make valuable manure. In colder places, the incineration of trash warms homes and working places. The Japanese have successfully converted paper-dominated waste into toilet paper. The slag produced in the making of steel (600 kg slag per tonne of steel produced) is granulated, powdered and then used in portland cements. The flyash generated while burning coal—in boilers to produce electricity, and in railway engines—is used as cementing material on a large scale.

These are a few of the many ways in which solid waste can be converted into resources.

# 19

# Finding Causes of Diseases: Medical Geology

## Trace Elements and Health

A large number of elements that are necessary for the well-being of humans and animals (and creatures that have no biological role in metabolism), occur variably in soils and dissolved in groundwater. But, they occur in trace amounts derived from the weathering of rocks. By direct consumption, or ingested through plants, these elements reach the bodies of humans and animals (Figure 19.1). Among the essential trace elements are calcium, chlorine, magnesium, lead, potassium, sodium, sulphur, chromium, cobalt, copper, fluorine, iron, manganese, molybdenum, zinc and selenium. Unwanted elements include cadmium, arsenic, mercury, lead and radon. In the matter of essential elements, there is a **safe** and **adequate range** of each element (Table 19.1). If the body takes less or more than the safe range of exposure, body cells may get impaired or depressed, leading to health problems including serious illnesses.

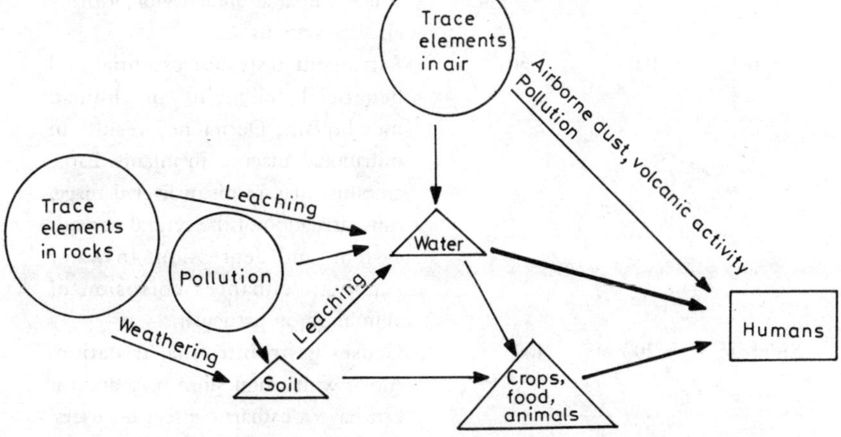


**Figure 19.1.** Various pathways through which trace elements enter the human body (After Montgomery, 1995).

**Table 19.1.** Safe and adequate range of trace elements, and the consequence in case of deficiency or excess in their micro-quantities (After Nautiyal et al, 2002).

| Parameters | BIS Prescribed Limits (mg/l) | | Probable effects when beyond permissible limit |
|---|---|---|---|
| | Desirable | Permissible | |
| Iron | 0.30 | 1.0 | Gives bitter-sweet astringent taste; causes staining of laundry and porcelain. In traces it is essential for nutrition. |
| Chloride | 250 | 1,000 | May be injurous to people suffering from diseases of the heart or kidneys. Causes indigestion and corrosion; pallate affected. |
| Residual chlorine | 0.20 | — | Excessive chlorination of drinking water may cause asthma, colitis and eczema. |
| Calcium | 75 | 200 | Causes encrustation in the water-supply system. Insufficiency causes a severe type of rickets, while excess causes concretions in organs such as the kidney or bladder (stones) and irritation in urinary passages (Essential for nervous and muscular systems, cardiac functions and coagulation of blood.) |
| Magnesium | 30 | 100 | Its salts are cathartic and diuretic. High concentration may have a laxative effect, particularly on new users. Magnesium deficiency is associated with structural and functional changes. It is essential as an activator of many enzyme systems. |
| Copper | 0.05 | 1.50 | Astringent taste but essential and beneficial element in human metabolism. Deficiency results in nutritional anaemia in infants. Large amounts may result in liver damage, cause irritation of the central nervous system, and depression. In water supply, it enhances corrosion of aluminium in particular. |
| Sulphate | 200 | 400 | Causes gastrointestinal irritation. Along with magnesium or sodium it can have a cathartic effect on users. More than 750 mg/l concentration may have laxative effects if given along with magnesium. |

| Parameters | BIS Prescribed Limits (mg/l) | | Probable effects when beyond permissible limit |
|---|---|---|---|
| | Desirable | Permissible | |
| Nitrate | 45 | 100 | Causes methaemoglobinaemin at very high concentration, causes gastric cancer and affects the central nervous system and cardiovascular system adversely. |
| Fluoride | 1.0 | 1.5 | Reduces dental caries; very high concentration may cause crippling skeletal fluorosis. |
| Cadmium | 0.01 | — | Acute toxicity may be associated with renal, arterial hypertension, itia-itia disease. Cadmium salt causes cramps, nausea, vomiting and diarrhoea. |
| Lead | 0.05 | — | Toxic in both acute and chronic exposures. Burning in mouth, severe inflammation of the gastrointestinal tract with vomiting and diarrhoea. Chronic toxicity produces nausea, severe abdominal pain, paralysis, mental confusion, visual disturbances, anaemia, etc. |
| Zinc | 5 | 15 | An essential and beneficial element in human metabolism. Taste threshold for zinc occurs at about 5 mg/l, imparts astringent taste to water. |
| Chromium | 0.05 | — | Hexavalent state of chromium produces lung tumours, can produce cutaneous and nasal mucous membrane ulcers and dermatitis. |
| Arsenic | 0.05 | — | Skin damage, circulatory system problems, increased risk of cancer. |
| Antimony | 0.006 | — | Increase in blood cholesterol, decrease in blood glucose. |
| Aluminium | 0.03 | 0.2 | Causes neurological disorder. |
| Barium | 2.0 | — | Increases blood pressure. |
| Beryllium | — | 0.0002 | Is carcinogenic. |
| Cyanide | 0.05 | — | Causes nerve damage, thyroid problems. |
| Mercury | 0.001 | — | Causes neurological and renal disturbances. Excess causes gonadotoxic and mutagenic effects and disturbs the cholesterol metabolism. |
| Boron | 1.0 | 5.0 | Affects central nervous system, its salts may cause nausea, cramps, convulsions, coma, etc. |

| Parameters | BIS Prescribed Limits (mg/l) | | Probable effects when beyond permissible limit |
|---|---|---|---|
| | Desirable | Permissible | |
| Manganese | 0.1 | 0.3 | Essential as a co-factor in enzyme systems and metabolic processes. Excessive intake causes change in appetite and reduction in metabolism of iron to form haemoglobin. Imparts undesirable taste and stains plumbing fixtures and laundry. |
| Selenium | 0.01 | 0.3 | Hair and fingernail loss, numbness in fingers and toes. Circulatory problems. |
| Calcium carbonate | 200 | 600 | Imparts distinctly unpleasant taste; may be deleterious to humans in presence of high pH, hardness and total dissolved solids. |
| Sodium | No guideline | | Harmful to persons suffering from cardiac, renal and circulatory diseases. |
| Potassium | No guideline | | An essential nutritional element but is laxative in excess. |
| Nickel | No guideline | | Non-toxic element but may be carcinogenic; can react with DNA resulting in DNA change. |

A deficiency in iodine causes enlargement of the thyroid gland or goitre. Excess intake through plants, via phosphate fertilizers or pyritic-carbonaceous shales, may result in accumulation of cadmium in the renal cortex, with consequent kidney lesions (wounds). High levels of phosphorus may induce calcium deficiency in bones, causing bone impairment. On the other hand, lack of phosphorus may lead to a tendency to eat substances other than food (pica), poor growth, infertility, or even wasting of bones or osteodystrophy (Table 19.2). Southern Kakadu, a region in Australia, is a place where people (both settlers and aborigines) dread to go, because it leads to strange sicknesses. There are localized areas where the natural levels of thorium, uranium, mercury, fluorine and radon are unusually high (Selinus, 2002).

**Table 19.2.** Diseases at states of deficiency and toxicity, respectively, caused by some elements (Selinus, 2002).

| Element | Deficiency | Toxicity |
|---|---|---|
| Iron | Anaemia | Haemochromatosis |
| Copper | Anaemia | Chronic copper poisoning |
| | Sway back | Wilson-, Bedlington disease |
| Zinc | Dwarfed growth | Metallic fever |
| | Retarded development of gonads | Diarrhoea |
| | Akrodermatitis enteropathica | |

| Element | Deficiency | Toxicity |
|---|---|---|
| Cobalt | Anaemia<br>White liver disease | Heart failure<br>Polycythaemia |
| Magnesium | Dysfunction of gonads<br>Convulsions<br>Malformations of skeleton<br>Urolithiasis | Ataxia (loss of full control of bodily movements) |
| Chromium | Disturbances in glucose metabolism | Kidney damage (nephritis) |
| Selenium | Liver necrosis<br>Muscular dystrophy (White muscle disease) | Alkali disease<br>Blind staggers |

## Medical Geology

Human activities could result in the redistribution of trace elements in soils and groundwater. Consequently, their concentration in some sites could become excessive or deficient—so that they have adverse effects on the quality of life. Knowledge of geochemistry and geohydrology helps geologists to identify and map sites where the presence of trace elements (within bedrocks, soils or groundwater) is not in keeping with the desirable level, where it is above or below the safe and adequate range. This is an exercise in medical geology, which 'deals with the influence of ordinary environmental factors on the geographic distribution of essential and toxic elements and consequent problems on humans and animals' (Selinus, 2002).

## Bioavailability of Trace Elements

Since most metals find their way into plants, particularly into the roots of aquatic vegetation and mosses, it is necessary to know their presence in different forms in various parts of the plant body. Understanding the **bioavailability of metals** is therefore an essential element in the practice of medical geology. The bioavailability of trace elements depends on environmental conditions and the nature of plants. Both the bioavailability and the mobility of zinc, lead and cadmium are greatest under acidic conditions in soils and water, but decrease with increasing pH values (Selinus, 2002). In many places, the above-ground parts of plants do not show concentration of metal, owing to physiological barriers that prevent the intake of toxic levels of metals. Roots, especially of aquatic plants, are therefore the most important part to be monitored for their chemistry. Mosses, likewise, closely respond to chemical variations in the background levels of different bedrocks and/or soils.

## Some Examples of Health Problems

### CARDIOVASCULAR DISEASES

In some stretches of the Kerala coast, underprivileged children suffer from a strange malady—emaciated legs, protruding bellies, puffy faces, necks with engorged veins, and shortness of breath. The symptoms are caused by the growth of a leathery-fibrous layer inside the pumping chambers of the heart, resulting from a high level of thorium in the tissues (Valiathan, 1998). The larger concentration of thorium is attributed to the mineral monazite that abounds in Kerala's coastal sands. Coming into contact with the monazite grains (with 8% thorium and ~ 30% cerium) in the coastal sands, and eating tuber crops like tapioca and elephant-foot yam both of which show a high ingestion of thorium, children who suffer from magnesium deficiency fall victim to the heart-muscle disease *endomycocardia fibrosis*. Monazite–ilmenite-rich sands occur in many places along the eastern and western coasts of India.

A new source of danger from natural causes has recently been perceived. The western side of Hyderabad, situated on granite that is rich in uranium (15–35 ppm) and thorium (60–150 ppm), receives a dose of adsorbed gamma rays that is two to five times higher than normal—at a rate of more than 350 millirad/year, and locally as high as 500–600 millirad/year (Srinivasan et al, 2001). This implies that the fashionable western part of the metropolis is exposed to the hazard of radioactive gamma rays.

Studies carried out in recent years in North America, Europe and Japan (Foster, 1992) have shown a significant relationship between the hardness of water and incidence of heart diseases—the lower the hardness, the higher the incidence of ischaemic heart disease. Hard water is a result of the dissolution of carbonates, dominantly of calcium and magnesium—it is related to the presence of carbonates and bicarbonates in a dissolved state, derived from limestones and dolomites. In fifteen countries in Europe, including the United Kingdom and Sweden, a high incidence of ischaemic heart disease was found in cities and towns that had soft water rather than hard water (Nerbrandt et al, 1992).

## ZINC AND DIABETES

The content of zinc is known to have direct bearing on the health of people, especially when subjected to long exposure to this element through drinking water. A low concentration of zinc is associated with 'later development of childhood onset of diabetes' (Haglund et al, 1996). The onset of childhood diabetes, it may be pointed out, is wholly dependent on auto-immune insulin. Thus, the higher the concentration of zinc, the lesser the risk of this disease.

## SELENIUM DEFICIENCY AND TOXICITY

Selenium is an essential element, important to the health of animals and humans. Its deficiency causes swelling of joints, pain and general malaise in the early stage, and crippling skeletal deformation in the advanced stage, as witnessed in a wide NE–SW belt in China, and in the Cossack country in Russia (Selinus, 2002). In northeastern parts of Panjab, embracing the districts of Hoshiarpur and Nawashahar, there are areas where excessive selenium manifests itself in the sugarcane crops, some vegetables and in cattle. The cattle show splitting of hair, change of nail colour, and deep cracks in hooves (Nautiyal et al, 2002). Among people there are cases of giddiness, lassitude, depression and garlic-tinged odour in breath. The groundwater is characterized by selenium levels ranging from 0.002 to 0.015 mg/l—in one case as high as 0.114 mg/l—which is much more than the critical level of 0.01 mg/l (Nautiyal et al, 2002). Another survey (Dhillon et al, 1992) recorded selenium content ranging from 2.54 to 69.53 microgram/litre in the seleniferous belt, and from 0.25 to 8.63 microgram/litre in the non-seleniferous areas. This explains the problem that people and their cattle are facing in these regions.

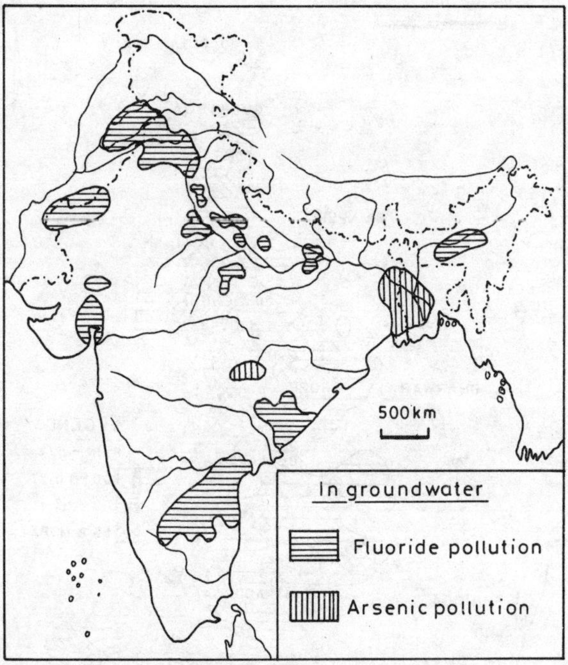


**Figure 19.2.** High content of fluorine and arsenic in the groundwater of several affected areas in India (Based on Jha and Jha, 1982).

## FLUORINE AND FLUOROSIS

Fluorine is an essential element in the development of bones, the recommended daily intake being 1.5–4 mg/l. As already pointed out in Chapter 5, the fluorine content in groundwater in many parts of Karnataka, Tamilnadu, Andhra Pradesh, Gujarat, Rajasthan, Madhya Pradesh, Uttar Pradesh and Panjab is very high (Figure 19.2). Fluorine pollution of groundwater has become a scourge for 62 million people living in 17 states. There are 9,741 villages and 6,819 habitations where people consume water having more than the admissible (1.5 mg/l) limit of fluorine.

The presence of fluorine in water (in excess of 1.5 mg/l) causes a crippling disease called *fluorosis*. It is manifest in the form of mottling and pitting of teeth (as in periosteum), and as calcification of ligament tendons, leading to deformation and eventual immobilization. This has happened to half a million people in southern India. In western Rajasthan, more than 3,500 people consume water that has more than 10 mg/l of fluorine (Agrawal et al, 1999)—in three villages of Sirohi district, dugwell and tubewell water has more than 16 ppm of this element (Maithani et al, 1998). In South India (Figure 19.3), in the Gulbarga–Raichur–Bellary belt, the groundwater contains between 7.4–8.5 mg/l of fluorine (Suma Latha et al, 1999; Susheela, 1999). In the Kolar area, where 26,000 people suffer from dental and skeletal fluorosis and 39,000 people are prone to it—including 80 per cent of children in the age group 6–14 years—the fluoride concentration in the groundwater is between 2.8–4.3 mg/l.

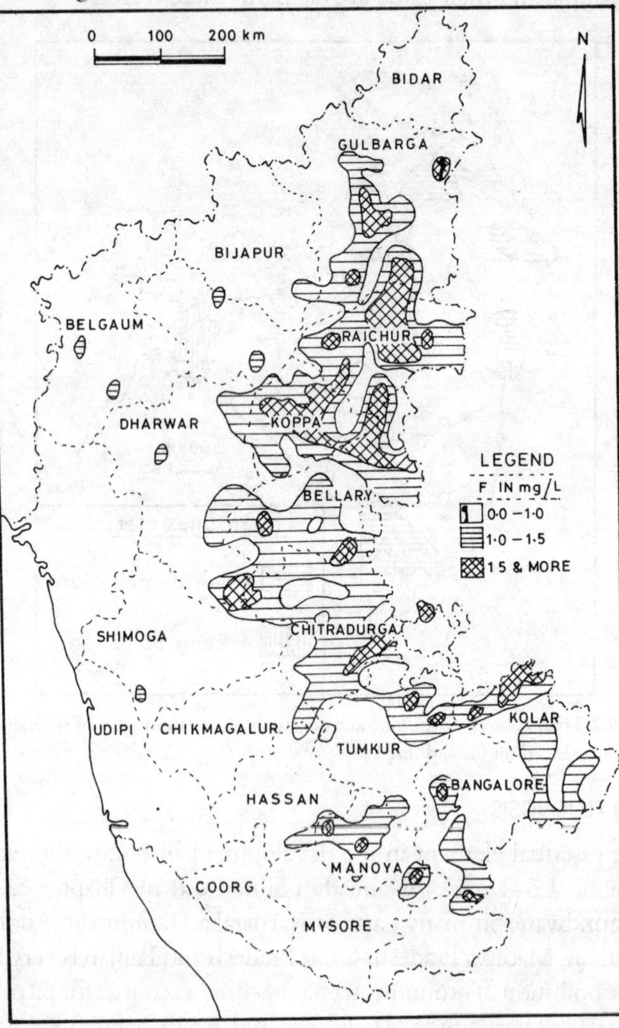


**Figure 19.3.** Distribution of fluorine in the groundwater below the water table in Karnataka (Suma Latha et al., 1999).

## ARSENIC POISONING

Arsenic poisoning of groundwater has assumed dangerous proportions in Bangladesh and West Bengal (Figure 19.4), where more than 600,000 people in about 400 villages have been disfigured and are terminally ill. Hundreds of wells supply water that contains 20–70 times more arsenic than the prescribed safe limits.

Significantly, neither surface water nor shallow and deep well water are subject to arsenic contamination. It is only water drawn from a 50–80 m deep aquifer that is poisonous. This situation is attributed to over-withdrawal of groundwater, and the resultant frequent fluctuation in the water table. Over-withdrawal has exposed the 50–80 m deep aquifer, containing arsenic, to oxidation. The arsenic leaches out through carriers of iron oxyhydroxides, derived from a ferruginous coating around sand grains (Mallick and Rajagopal, 1996). This has happened east of the Bhagirathi (Ganga) river, up to the Padma river plain, in 8,000–25,000-year-old sediments (Figure 19.4). In the Baruipur block south of Kolkata, iron oxyhydroxide coating on sand, and siderite concretion in iron pyrite-bearing layers (at depths), are responsible for the release of the poison to the aquifer. This follows the oxidation of arsenic-bearing sulphide, aided by microbial activity, as borne out by the enrichment of organic matter in the arsenic (Acharyya et al, 2000, Pal et al, 2002). In the deltaic area of the Meghna river in Bangladesh, arsenic poisoning is distributed upto a depth of 366 m (Anwar et al, 2002). Here too, the elevated level of arsenic is found predominantly in organic matter and in the iron oxyhydroxide coating on sand grains.

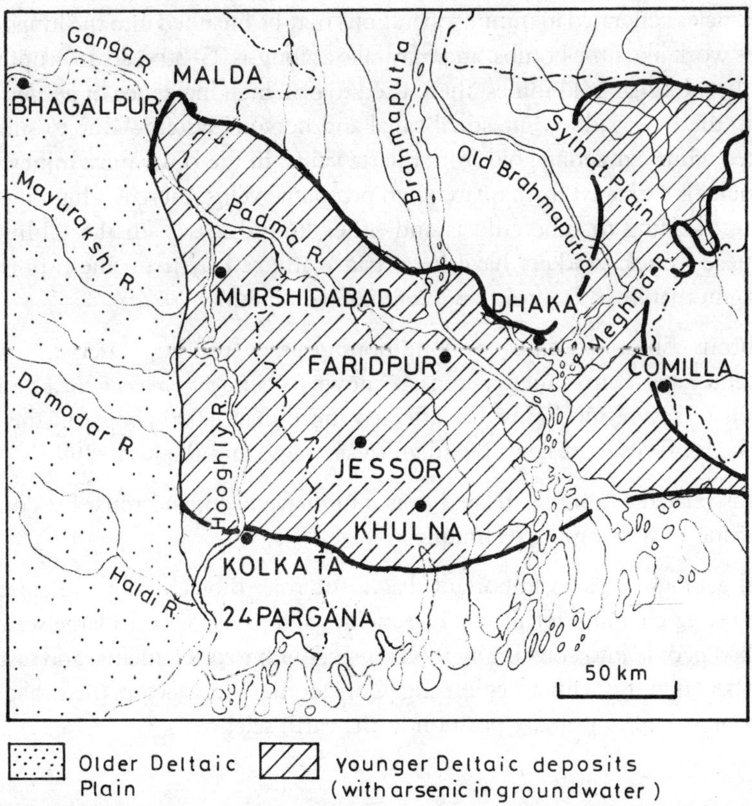


**Figure 19.4.** Arsenic poisoning in wells that tap a 50–80-m deep aquifer in the Younger Deltaic Deposit in the floodplain between the Bhagirathi and Padma rivers in West Bengal and Bangladesh (Based on Mallick and Rajagopal, 1998; Acharyya et al, 2000; Anwar et al, 2002).

## Tempering Toxicity of Trace Elements

There are some trace elements, the presence of which may have tempering effects on the poisonous qualities of other elements. In other words, the effects of a particular element is often modified in the presence of other elements (Selinus, 2002). It is well-known in Europe that sheep and cattle suffering from an increased intake of molybdenum need a greater amount of copper in their feed. It is also realized that the deficiency of one element may upset the balance of another. For example, the presence of molybdenum, selenium and lead can reduce the negative effects of a high concentration of fluorine. High intake of zinc in plants and animals may provide protection against the harmful effects of cadmium. Selenium is known to counter the adverse effects of arsenic as well as some mercury compounds. The level of selenium reduces on applying sulphur-bearing fertilizers to the soil. Likewise, plants that grow in phosphorous-rich soil suffer from deficiency in zinc. The plant fibre may inhibit the absorption of zinc, thus causing zinc deficiency in plants, and in the animals that consume it.

## Mining-related Ailments

Dust particles generated in mining operations may be breathed into the lungs. These particles work like time bombs, and may take as long as 25–30 years to erupt into an illness. In the Kolar gold mines, there used to be a high incidence of *pneumoconiosis* affecting the liver, kidney, intestinal canal and nervous system—due to silica dust produced while extracting gold from quartz veins. In the Bihar mica mines, where the mineral is extracted from quartz-rich pegmatite veins, there is a high incidence of *silicosis*, leading to tuberculosis and other lung diseases. In the Chhindwara manganese mines, workers have been badly affected by manganese poisoning, resulting in mania, pathogenic laughter and even insanity.

The fibrous asbestos used as air conditioning pipes, roofing material, sanitation fittings and water-conveyance systems, is known to cause *pneumoconiosis*. This illness, that leads to severe morbidity, has affected more than 2,000 people in the Jharal, Deogarh and Beawar areas in Rajasthan (Mansingka and Ranawat, 1996).

Problems related to mining-related poisoning have been, and must be, contained by resorting to wet-mining techniques.

Medical geology plays an important diagnostic role. Biotechnology in agriculture and the use of chemical fertilizers, insecticides and pesticides on a large scale, have compelled people into consuming foods tinged with toxic elements; and fast foods and snacks often use artificial colouring. Considering these factors medical geology could soon occupy a primary position in the earth sciences.

## References

Abbas, N., and V. Subramanian. 1984. Erosion and sediment transport in the Ganga basin. *Ind. Jour. Hydrology* 69: 173–182.

Abrol, I.P. 1994. Land degradation: A challenge to sustainability. In *Salinity management for sustainable agriculture,* ed. D.L.N. Rao, N.T. Singh and R.K. Gupta. Karnal: Central Soil Research Institute. VII–VIII.

Acharyya, A., B. Sengupta, and R. N. Ghosh. 2000. Mid-1997 tidal surge of Bay of Bengal at Digha on the east coast of India: Its damage evaluation and mitigation. *Jour. Nepal Geol. Soc.* 21: 75–86.

Agarwal, Anil, Sunita Narain, and Srabani Sen, eds. 1999. *The citizens' fifth report* (2 parts). New Delhi: Centre for Science & Environment. 440 pp. and 256 pp.

Agarwal, R.P., J.N. Rai, and A. Kumar. 1996. Geoenvironmentally compatible management of groundwater in Sharada Sahayak Canal command area in parts of Mau district, U.P. In *Proc. Special Publication No. 48(i)*. Lucknow: Geological Survey of India. 207–211.

Agarwal, S.K., 2001. Mining industry in India facing global challenges. *Minerals & Metals Review* (November): 12–13.

Ahmad, M., S. Ratan, and B.B. Dhar. 1996. The impact of processing of lead-zinc ores on the surrounding environment. In *The impact of mining on the environment*. Rotterdam: A.A. Balkema. 229–236.

Anbalagan, R., and B. Singh. 2001. Landslide hazard and risk mapping in the Himalaya. In *Landslide hazard mitigation in the Hindukush-Himalaya*, ed. Li Tianchi, et al. Kathmandu: ICIMOD. 163–188.

Anonymous. 1999. *Indian mineral year book 1998–1999*. Nagpur: Indian Bureau of Mines. 668 pp.

Anonymous. 2000a. *State of the forest report 1999*. Dehradun: Forest Survey of India. 113 pp.

Anonymous. 2000b. *Reclamation/restoration techniques and strategy for mined areas*. Nagpur: Indian Bureau of Mines. 122 pp.

Anonymous. 2001. *Statistical outline of India 2001–2002*. Mumbai: Tata Services. 298 pp.

Ansari, A.A., I.B. Singh, and H.J. Tobschall. 1999. Status of anthropogenically induced metal pollution in the Kanpur–Unnao industrial region of the Ganga plains. *Environ. Geol.* 38: 25–33.

Arendt, A.A., K.A. Echelmeyer, W.D. Harrison, C.S. Lingle, and U.B. Valantine. 2002. Rapid wastage of Alaska glaciers and their contributions to rising sea level. *Science* 297: 382–386.

Arya, A.S. 2000. Recent developments towards earthquake risk reduction. *Current Science* 79: 1270–1277.

———. 2001. Case study in India: Diagnosis, repair and strengthening of damaged buildings. In *Disaster management*, ed. V.K. Sharma. New Delhi: Indian Institute of Public Administration. 157–176.

Baba, M. 1979. Beach erosion in Kerala: Problems and some solutions. *Professional Papers No. 6*. Trivandrum: Centre for Earth Science Studies. 15 pp.

Babu, R., K.G. Tejwani, M.C. Agarwal, and L.S. Bhushan. 1977. Distributions of erosion index and isoerodant map of India. *Indian Jour. Soil Conservation* 6: 1–12.

Bartarya, S.K., and K.S. Valdiya. 1989. Diminishing discharges of mountain springs in a part of Kumaun Himalaya. *Current Science* 58: 417–426.

———, N.S. Virdi, and M.P. Sah. 1996. Landslide hazards: Some case studies from the Satluj valley, H.P. *Himalayan Geology* 17: 193–207.

Barua, D.K., 1997. The characteristics and mobility of some major rivers in Bangladesh. *Water Nepal* 5: 109–127.

Bhan, Suraj, and Shamsher Singh. 1994. Soil and water conservation scenario in India. In *Soil and water conservation* (Vol. 2), ed. L.S. Bhushan, I.P. Abrol, and M.S.R. Mohan Rao. Dehradun: Indian Association for Soil & Water Conservation. 883–901.

Bhandari, R.K. 2002. *Seismic microzonation* (unpublished report). Chennai: Centre for Disaster Mitigation and Management, Anna University. 198 pp.

Bhutiyani, M.R. 1992. The avalanche problems in Nubra and Shyok valleys in Karakoram Himalaya. *Jour. Inst. Military Engineering* 3(3): 3–5.

———, V.S. Kale, and N.J. Pawar. 2002. Variations in glaciohydrological characteristics of some southwestern Himalayan river basins in this century. In *Geodynamics and Environment Management of Himalaya*, ed. H.C. Nainwal and C. Prasad. Shrinagar: H.N.B. Garhwal University. 130–138.

Bilham, R., F. Blume, R. Benedick, and V.K. Gaur. 1998. Geodetic constraints on the transition and deformation of India: Implications for future great Himalayan earthquakes. *Current Science* 74: 213–229.

———, and V.K. Gaur. 2000. Geodetic contributions to the study of seismotectonics in India. *Current Science* 79: 1259–1268.

Bisht, K.S., and M.P. Sah. 1999. The devastating landslide of August 1998 in Ukhimath area, Rudraprayag district, Garhwal Himalaya. *Current Science* 76: 481–484.

Borcherdt, R.D., ed. 1975. Studies for Seismic Zonation of the San Francisco Bay Region. Washington: U.S. Geol. Survey Professional Paper 941A.

Bradshaw, A.D. 1997. The importance of soil ecology in restoration science. In *Restoration ecology and sustainable development,* ed. K.M. Urbanska, et al. Cambridge: Cambridge University Press. 33–64.

———, and M.J. Chadwick. 1980. *The restoration of derelict land.* Oxford: Blackwell. 317 pp.

Brown, Jennifer, ed. 1991. *Environmental threats: Perception, analysis and management*. New Delhi: CBS Publications. 154 pp.

Brown, L.R. 2001. *Eco-economy: Building an economy for the earth*. Hyderabad: Orient Longman. 334 pp.

Cardona, C., L. Dwelley, S. Trumbull, B.E. Tucker, and C. Villacis. 1998. Improving urban earthquake risk management in developing countries. In *Proc. Intern. Sym. Seismic Risk in Megacities*. Trieste: ICTP. 1–3.

Chaddha, D.K. 1999. Groundwater. In *Vision 2050*. New Delhi: Indian Water Resources Society. 8–14.

Chandra, P.C., A. Ramakrishna, S.C. Singh, and P.H.P Reddy. 2002. Geophysical approach to delineate natural fracture zones in hard rocks. In *Sustainable development and management of groundwater resources in semi-arid regions with special reference to hard rocks*, ed. Thangarajan, et al. New Delhi: Oxford. 75 81.

Chaplow, R. 1981. Investigation of hydroelectric projects. In *Proc. Sem. Accelerated Hydroelectric Development in India*. New Delhi: Central Board of Irrigation & Power. 9–16.

Chaturvedi, R.S., D.C. Bhattacharya, P. Kamal, J. Krishnamurthy, and S.N. Raman. 1983. Integrating remote sensing techniques in groundwater exploration—a typical case study from Bundelkhand region. In *Proc. Nat. Sym. Remote Sensing in Development and Management of Water Resources*. Dehradun: Indian Institute of Photointerpretation & Remote Sensing. 267–276.

———, R.S. Sharma, and A.K. Tangri. 1983. Evaluation of hydrological regions in Saryu Project area of Gonda and Bahraich districts, U.P. In *Proc. Nat. Symp. Remote Sensing in Development and Management of Water Resources*. Dehradun: Ind. Institute of Photointerpretation & Remote Sensing. 113–117.

Chen, W.P., and P. Molnar. 1977. Seismic moments of major earthquakes and the average rate of slip in central Asia. *Jour. Geophys. Res.* 82: 2945–2969.

Dasgupta, A.B. 1999. Estimates of hydrocarbon resources and reserves, and petroleum exploration and production scenario of India: An overview. *Bulletin Oil & Natural Gas Corporation* 36: 29–34.

Deng, Q., Pu Jiang, L.M. Jones, and P. Molnar. 1981. A preliminary analysis of reported changes in groundwater and anomalous behaviour before the 4 February 1975 Haicheng earthquake. In *Earthquake prediction: An international review*. Washington: Amer. Geophys. Union. 553–565.

Deodhar, N.S., and K.J. Nath. 1981. Water pollution and its control in India. In *Impact of development of science and technology on environment*, ed. A.K. Sharma and Archana Sharma. Kolkata: Ind. Science Congress Assoc. 152–161.

Deshmukh, Megha. 1995. Ralegaon-Sidri watershed, a case study of artificial recharge of groundwater. In *Artificial recharge of groundwater*, ed. A.I. Johnson and D.G. Payne. New York: Amer. Soc. Civil Engineers. 839–847.

Dhar, R.K., B.K. Biswas, G. Samanta, and 4 others. 1997. Groundwater arsenic calamity in Bangladesh. *Current Science* 73: 48–59.

Dhruva Narayana, V.V. 1993. *Soil and water conservation research in India*. New Delhi: Indian Council of Agricultural Research. 448 pp.

———, G. Sastry, and U.S. Patnaik. 1997. *Watershed management*. New Delhi: Indian Council of Agricultural Research. 176 pp.

Dhillon, K.S., S.K. Dhillon, and S.S. Bawa. 1992. Selenium toxicity in some plants and soils in Punjab. *Jour. Ind. Soc. Soil Sci.* 40: 132–136.

Dillon, P. 2002. Banking of storm water, reclaimed water and potable water in aquifers. In *Sustainable development and management of groundwater resources in semi-arid regions with special reference to hard rocks*, ed. Thangarajan, et al. New Delhi: Oxford. 129–137.

Dube, S.K., and Gaur, V.K. 1995. A PC-based operational storm surge prediction system for disaster management in coastal India. *Current Science* 68: 1103–1113.

Duraiswami, R.A., S.S. Thigale, N.J. Pawar, and D.C. Meshram. 2000. Coastal environment of Maharashtra: Issues vis-à-vis development. In *Contributions to environmental geosciences,* ed. A.M. Pathan and S.S. Thigale. New Delhi: Aravali Books. 59–73.

Foster, H.D. 1992. *Health, disease and the environment*. Boca Raton: CRC Press. 614 pp.

Freymueller, J., R. Bilham, R. Burgmann, K.M. Larson, J. Paul, S. Jade, and V.K. Gaur. 1996. Global positioning system movements of the Indian plate: Motion and convergence across the Lesser Himalaya. *Geophys. Res. Letters* 23: 3107–3110.

Galay, V.J., T. Okaji, and K. Nishino. 1995. Erosion from the Kulekhani watershed, Nepal, during the July 1993 rainstorm. In *Challenges in mountain resource management in Nepal*, ed. H. Schreier, P.B. Shah, and S. Brown. Kathmandu: ICIMOD. 13–24.

Garg, J.K., K.G. Saxena, A. Narayana, T.V.R. Murthy, R.K. Maikhuri, A.S. Arya, V. Joshi, K.S. Rao, V.P. Sati, and S. Ghosh. 1995. *Environmental appraisal of Tehri dam multipurpose river valley project using remote sensing and GIS*. Almora: G.B. Pant Institute of Himalayan Environment, and Ahmadabad: Space Applications Centre (ISRO). 119 pp.

Garg, S., S.V. Veeravalli, and S.R. Kale. 1999. Dams—an engineering analysis of alternatives. *Ind. Jour. History of Science* 34(4): 289–311.

Gaur, V.K. 1992. Earthquake hazard in the Himalaya. In *Earthquake hazard and large dams in the Himalaya*, ed. V.K. Gaur. New Delhi: Indian National Trust for Art and Cultural Heritage. 63–74.

Ghassemi, F., A.J. Jakeman, and H.A. Nix. 1995. *Salinization of Land and Water Resources*. Canberra: Centre for Resources and Environmental Studies, Australian National University. 335 pp.

Ghose, A.K. 1982. Environmental control and mining industry. *Jour. Mines, Metals and Fuels* 30: 167 pp.

Ghosh, R., and T.S. Bathija. 2001. Need for speedy water resources development in India. *Ind. Jour. Power & River Valley Development* (Golden Jubilee Vol.): 73–77.

Ghosh, Sutapa. 2003. Land subsides in groundwater guzzling districts of U.P. *Down to Earth* (March 31): 7–8.

Goswami, D.C. 1998. Fluvial regime and flood hydrology of the Brahmaputra river, Assam. In *Flood studies in India*, ed. V.S. Kale. Bangalore: Geological Society of India. 53–76.

Goswami, S. 1998. Assam's dire need. In *Souvenir, National workshop on geodynamics of northeastern India.* Dibrugarh: Dibrugarh University. 36–41.

Grimshaw, R.G., and L. Helfer. 1995. *Vetiver grass for soil and water conservation, land rehabilitation and embankment stabilization.* Washington: The World Bank. 281 pp.

Guha, S.K. 2000. *Induced earthquakes*. Dordrecht: Kluwer Academic Publishers. 314 pp.

Gupta, G.P. 1975. Sediment production—Status report on data collection and utilization. *Soil Conservation Digest* 3(2): 10–21.

Gupta, H.K. 1992. *Reservoir-induced earthquakes.* Amsterdam: Elsevier. 470 pp.

Gupta, J.P. 1975. Evaluation of the recharge techniques adopted for conjunctive use of surface and groundwater in the state of Haryana. In *Artificial recharge of groundwater II*, ed. A.I. Johnson and D.G. Payne. New York: Amer. Soc. Civil Engineers. 455–463.

Gupta, M.C., V.K. Sharma, L.C. Gupta, and B.K. Tamini. 2001. *Manual on natural disaster management in India*. New Delhi: Department of Agriculture & Cooperation, Government of India. 204 pp.

Gupta, Ritu. 2002. Lull before the storm. *Down to Earth* (April 15): 22–23.

Gupta, R.P., A.K. Saha, M.K. Arora, and A. Kumar. 1999. Landslide hazard zonation in a part of the Bhagirathi Valley, Garhwal Himalaya, using integrated remote sensing GIS. *Himalayan Geology* 20: 71–85.

Haglund, B., K. Ryckenberg, O. Selinus, and G. Dahlqvist. 1996. Evidence of a relationship between childhood-onset type I diabetes and low groundwater concentration of zinc. *Diabetes Care* 19: 873–875.

Haigh, M.J. 2000a. Landslide prediction: criticality magnitude and frequency. In *Proc. Intern. Workshop Geoenvironmental Hazards*, ed. V. Subramanian and R. Jayakumar. UNESCO-Envis Centre. 1–14.

———. 2000b. Headwater control: Despatches from the research front. In *Environmental*

*Reconstruction In Headwater Areas*, ed. M. Haigh and J. Krecek. Drodrecht: Kluwer Academic Publications. 25–52.

Haigh, Martin, and Josef Krecek. 2000. *Environmental reconstruction in headwater areas*. Drodrecht: Kluwer Academic Publications. 242 pp.

Harvey, C.F., C.H. Swartz, A.B.M. Badruzzaman, and 12 others. 2002. Arsenic mobility and groundwater extraction in Bangladesh. *Science* 298: 1602–1606.

Ikeya, M. and 6 others. 1997. Pulsed electric field before Kobe and Izu earthquakes from seismically induced anomalous animal behaviour. *Episodes* 20: 253–260.

Iyengar, R.N. 1998. Seismic risk status of Indian megacities. In *Proc. Intern. Sym. Seismic Risk in Megacities*. Trieste: ICTP. 10–11.

Jamwal, Nidhi. 2002. Swamped. *Down to Earth* (March 31): 34–38.

Jha, V.K. 1982. Highway alignment survey in mountainous areas and its impact on environment—a case example from Okhaldhunga area, Nepal. In *Fourth Intern. Congr. of Engineering Geol.* New Delhi: Oxford. 403–412.

Jha, L.B., and M. Jha. 1982. Fluoride pollution in India. *Intern. Jour. Environ. Studies* 19: 225–230.

Jones, L.M. 1995. *Pulling down roots in earthquake country*. Los Angeles: U.S. Geol. Survey, and Southern California Earthquake Centre. 29 pp.

Jones, L., and P. Molnar. 1976. Frequency of foreshocks. *Nature* 262: 677–679.

Joshi, D.D. 2001. Geoenvironmental problems of the glacial belt of Bhutan High Himalaya and remedial measures: A case study of Raphstreng Lake in Lunana area. *Jour. Geol. Soc. Ind.* 57: 65–76.

Juyal, G.P., K.S. Dadhwal, V.S. Katiyar, P. Joshie, and R.K. Arya. 1998. Mined area rehabilitation in Himalaya: Sahasradhara experience. In *Himalayan environment and sustainable development*, ed. P.B. Saxena. Dehradun: Surya Intern. Publication. 254–263.

Kale, Vishwas S. 1998. Monsoon floods in India: A hydrogeomorphic perspective. In *Flood studies in India*, ed. V.S. Kale. Bangalore: Geological Society of India. 229–256.

———. 1999. Long-period fluctuations in monsoonal floods in the Deccan Peninsula. India. *Jour. Geol. Soc. Ind.* 53: 5–15.

Karanth, K.R. 1997. On sustainable yield of infiltration gallery in valley fills of Peninsular India. *Jour. Applied Hydrology* 10: 86–98.

Karmarkar, B.M., S.R. Kulkarni, and S.S. Marathe. 1995. Basalt geology and recharge by percolation tanks. In *Artificial recharge of groundwater II*, ed. A.I. Johnson and D.G. Payne. New York: Amer. Soc. Civil Engineers. 742–750.

Kattlemann, R. 1987. Uncertainty in assessing Himalayan water resources. *Mountain Research & Development* 7: 279–286.

Kaul, R.N., and D.C. Das. 1982. Soil and water management. In *State of environment—some aspects*. New Delhi: Nat. Com. Planning Department of Environment & Forests. 1–16.

Kayastha, S.L. 2000. Water: Sustainable management of a vital resource. *Vasudha* 3: 1–36.

Keilis-Borok, V.I., T.L. Kronvod, and G.M. Molchen. 1984. Seismic hazard for largest cities of the world. In *Earthquakes and geological hazard prediction*, ed. S.A. Fedov. Moscow: Intern. Geol. Congr. 12–34.

Keller, E.A. 1976. *Environmental geology*. Columbus: Bell & Howell. 488 pp.

Khan, A.A., S. Hoque, S.M. Huq, K.Q. Kibria, and M.A. Hoque. 2003. Evidence of bacterial activity in the release of arsenic—a case study from the Bengal delta of Bangladesh. *Jour. Geol. Soc. Ind.* 61: 209–214.

Khan, M.A. 1998. Rainwater harvesting and recycling in western Rajasthan, India. In *Soil and water conservation: Challenges and opportunities* (Second edn.), ed. L.S. Bhushan, et al. New Delhi: Oxford. 404–412.

Khattri, K.N. 1995. Observed and theoretical acceleration response spectra in the Tehri region: Implication for seismic hazard in the regions. *Current Science* 69: 161–171.

———. 1999a. Probabilities of occurrence of great earthquakes in Himalaya. *Current Science* 77: 967–972.

———. 1999b. A seismic hazard and risk scenario for Himalaya and Ganga plains due to a future great earthquake. *Proc. Ind. Acad. Sci. (Earth & Planet. Sci)* 108: 87–92.

Kilmartin, M., and M.J. Haigh. 1998. Land reclamation policies and practices. In *Mining and environment in India*, ed. S.C. Joshi and G. Bhattacharya. Nainital: Himalayan Research Group. 467 pp.

Kimothi, M.M., and N. Juyal. 1996. Environmental impact assessment of a few selected watersheds of the Chamoli district (Central Himalaya) using remotely sensed data. *Intern. Jour. Remote Sensing* 17: 1391–1405.

Krynine, D.P., and W.R. Judd. 1957. *Principles of engineering geology and geotechnics.* New York: McGraw-Hill. 730 pp.

Kulkarni, R.P. 1982. Irrigation engineering in India. *Ind. Jour. History of Science* 17(1): 28–45.

Lagwankar, V.G., A.K. Garde, and K.D. Patil. 1994. Trends in reservoir sedimentation in India and abroad. In *Proc. Workshop on Reservoir Sedimentation*. Bangalore: Irrigation Department, Government of Karnataka. 127–134.

Lal, Murari, T. Nozawa, S. Emori, and 7 others. 2001. Future climate change: Implications for Indian summer monsoon and its variability. *Current Science* 81: 1196–1207.

Lavania, U.C. 2000. Vetiver grass technology for environmental protection and sustainable development. *Current Science* 78: 944–946.

Legget, R.F. 1962. *Geology and engineering.* New York: McGraw-Hill. 884 pp.

Leopold, Aldo. 1941. Lakes in relation to terrestrial life pattern. In *Proc. Hydrology*. Madison: Wisconsin University Press. 17–22.

Leverson, D. 1980. *Geology and urban environment*. New York: Oxford University Press. 120 pp.

Lohani, B.N. 1997. Environmental impact assessment: Arun III Hydroelectric Project, Nepal. In *Environmental impact assessment for developing countries in Asia* Vol. 2 (Case Studies), ed. B.N. Lohani and 5 others. Bangladesh: Asian Development Bank. 1–57.

———, J.W. Evans, R.R. Everitt, H. Ludwig, R.A. Carpenter, and Shih-Liang Tu. 1997. *Environmental Impact Assessment for Developing Countries in Asia*. Kathmandu: Asian Development Bank. 199 pp.

Madhavan, N., and V. Subramanian. 2000. Sulphide mining as a source of arsenic in the environment. *Current Science* 78: 702–708.

Maithani, P.B., R. Gurjar, R. Banerjee, B.K. Balaji, S. Ramachandran, and R. Singh. 1998. Anomalous fluorine in groundwater from western parts of Sirohi district, Rajasthan, and its crippling effects on human health, *Current Science* 74: 733–781.

Mallick, S., and N.R. Rajagopal. 1996. Groundwater development in the arsenic affected alluvial belt of West Bengal—some questions. *Current Science* 70: 956–958.

Mandal, B.K., T.R. Chowdhury, G. Samanta, and 10 others. 1996. Arsenic in groundwater in seven districts of West Bengal: The biggest arsenic calamity in the world. *Current Science* 70: 976–986.

Mande, O.D., and D.R. Vohra. 1988. Nathpa–Jhakri Project—a landmark in hydropower development in Himalayan region. In *Hydropower development in the Himalayas,* ed. V.D. Choubey. New Delhi: Oxford. 239–247.

Mansingka, B.K., and P.S. Ranawat. 1996. Mineral economics and occupation health hazards of the asbestos resource of Rajasthan. *Jour. Geol. Soc. Ind*. 47: 375–382.

Marsh, W.M., and J. Grossa Jr. 1996. *Environmental geography.* New York: John Wiley.

Masud, Basit 2001. Flash floods on the Indus river in Pakistan. In *Mountain flash floods*. Kathmandu: ICIMOD. 12–13.

Meher-Homji, V.M. 1996. Monsoon: A bio-climatologist's point of view. *Current Science* 71: 352–357.

Mehta, K.K. 1983. *Reclamation of alkali soils in India*. New Delhi: Oxford. 280 pp.

Mohan Rao, N., and N. Rangachari. 1983. Snow avalanches of Himalaya—an analysis. In *Proc. First Nat. Symp. Seasonal Snow Cover.* New Delhi: Vigyan Bhavan.

Molnar, P. 1990. A review of the seismicity and the rates of active thrusting and deformation at the Himalaya. *Jour. Himalayan Geology* 1: 131–154.

Montgomery, C.W. 1995. *Environmental geology.* New York: Wm. C. Brown Publishers. 496 pp.

Mool, Pradeep. 2001. Glacial lakes and glacial lake outburst flood events in the Hindukush–Himalaya region. In *Mountain flash floods*. Kathmandu: ICIMOD. 7–9.

Mukherjee, B.P., and C.V. Sangewar. 2001. Recession of Gangotri glacier through 20th century. In *Role of earthscience in integrated development and related societal issues.* Kolkata: Geol. Survey of India. 1–3.

Muralidharan, D. 2003. *Andhra Pradesh ke Chittor Zile mein peya jal ke sanchayan evam vitaran ka ek anutha prayog. Vasundhara* (2003): p. 21.

Murthy, Y.K., and B.S. Srinivasa. 1972. Influence of geology on dam design. In *Geology and engineering problems of river valley projects,* ed. V.S. Krishnaswamy and P.N. Mehla. Kolkata: Ind. Society of Engineering Geology. 1–8.

Naibo, Z. 1986. Cascade development and multipurpose utilization of small rivers. *Urja* 19(3): 179–184.

Narain, Hari. 1995. *Integrated development of natural resources* (unpublished report). Hyderabad: National Geophysical Research Institute. 23 pp.

Nautiyal, M.D., A. Kumar, B.P. Singh, and D.S. Saini. 2002. Impact of water quality on human health with special reference to selenium toxicity. (Preprint)

Nawani, P.C., R. Sanwal, and H.C. Khanduri. 1991. On disproval of river-bed dislocation at Tehri dam site in the light of recent exploration. *Jour. Engin. Geol*. 20: 98–107.

Nayak, S., R.K. Sarangi, and A.S. Rajawat. 2001. Application of IRS-P4 OCM data to study impact of cyclone on coastal environment of Orissa. *Current Science* 80: 1208–1213.

Negi, G.C.S., and V. Joshi. 1996. Geohydrology of springs in a mountain watershed: The need for problem-solving research. *Current Science* 71: 772–776.

———, ———, and K. Kumar. 1998. Spring sanctuary development to meet household water demand in the mountains: A call for action. In *Research for moutain developments: Some initiatives and accomplishments*. Nainital: Gyanoday Prakashan. 25–46.

Nerbrandt, C., K. Svardsudd, J. Ek., and G. Tibblin. 1992. Cardiovascular mortality and morbidity in seven countries in Sweden, in relation to water hardness and geological settings. *European Heart Journal* 13: 721–727.

Pal, Taraknath, P.K. Mukherjee, and S. Sengupta. 2002. Nature of arsenic pollutants in groundwater of Bengal basin—a case study from Baruipur area, West Bengal. *Current Science* 82: 554–560.

Pancholi, D.M. 1997. Engineering geology and seismotectonic investigation, Sardar Sarovar Project Gujarat. *Miscellaneous Publication No. 63*. Kolkata: Geological Survey of India. 95–101.

Pant, G., G.L. Narula, and S.K. Shome. 1980. *Geotechnical studies in relation to the planning and designing Tehri dam project* (unpublished report). Lucknow: Geological Survey of India.

Parasuram, S., and P.V. Unnikrishnan, eds. 1998. *India disaster report: Towards a policy initiative*. New Delhi: Oxford University Press. 384 pp.

Pathak, B.D. 1983. Groundwater resources of U.P., India. In *Engineering Geology*, ed. B.B. Singhal. Meerut: Sarita Prakashan. 111–131.

Patnaik, N. 1978. Soil conservation and watershed management for flood retardation and preventation. *Ind. Jour. Soil Conservation* 6: 43–55.

Paul, D.K. 2004. Public policies for hazard management: Appropriate structural designs and building codes. In *Coping with natural hazards: Indian context*, ed. K.S.Valdiya. Hyderabad: Orient Longman.

Paul, S.K., S.K. Bartarya, P. Rautela, and A.K. Mahajan. 2000. Catastrophic mass movement of 1998 monsoons at Malpa in Kali Valley Kumaun Himalaya. *Geomorphology* 35: 169–180.

Pearce, P. 1983. Landfilling: A long term option for hazardous waste disposal. *Industry & environment.* Paris: UNEP. (Special Issue No. 41983): 57–63.

Petak, W.J., and A.P. Atkisson. 1982. *Natural hazard risk assessment and public policy*. New York: Springer. 489 pp.

Phadtare, P.N., and E. Bradley. 1983. Optimum artificial recharge potential in the alluvial aquifers of the Mehsana areas Gujarat. In *Proc. Sem. Assessment Development and Management of Groundwater Resources*. New Delhi: Central Groundwater Board. 307–316.

Pisharoty, P.R. 1987. Water management and wasteland development. *Wasteland news*. New Delhi: Society for Promotion of Wasteland Development. 3: 3–4.

Prudhvi, Raju K.N., and S.V.B. Krishna Bhagwan. 1984. Human-induced geomorphic hazards in Visakhapatnam coastal strip, A.P. *Trans. Insti. Indian Geographers* 6(1): 79–84.

Rai, K.L. 2000. Coal mining environment and related geohazards: The Indian scenario. In *Contributions to environmental geosciences*, ed. A.M. Pathan and S.S. Thigale. New Delhi: Aravali Books. 221–228.

Rai, S.C., and E. Sharma. 1998. Comparative assessment of runoff charactristics under different landuse patterns within a Himalayan watershed. *Hydrological Processes* 12: 2235–2248.

Rai, S.P., K.S. Valdiya, and J.S. Rawat. 1998. Management of water resources: Spring sanctuaries. In *The Khulgad Project: An experiment in sustainable development*, ed K.S.Valdiya. Nainital: Gyanodaya Parakashan. 41–60.

Rajawat, A.S., C.V.S. Sastry, and A. Narain. 1999. Application of pyramidal processing on high resolution IRS/IC data for tracing migration of the Saraswati river in parts of Thar Desert. In *Vedic Saraswati: Evolutionary history of a lost river in northwest India*, ed. B.P.Radhakrishna and S.S.Merh. Bangalore: Geol. Society of India. 259–272.

Rajendran, C.P. 2000. Using geological data for earthquake studies: A perspective from Peninsular India. *Current Science* 79: 1251–1258.

Rajendran, Kusala, and C.M. Harish. 2000. Mechanism of triggered seismicity at Koyna: An evaluation based on relocated earthquakes. *Current Science* 79: 358–368.

Ramachandra Rao, M.B. 1953. *A compilation of papers on the Assam earthquake of August 1950*. New Delhi: Central Board of Geophysics, Govt. of India. 112 pp.

Ramalingeswara Rao, B. 2000. Historical seismicity and deformation rates in the Indian Peninsular Shield. *Jour. Seism*. 4: 247–258.

Ramachandran, B. and S. Gangopadhyay. 1972. Engineering geological features of soft soil areas. *Record Geol. Survey India*. 104(2): 65–84.

Ramanathan, V., P.J. Crutzen, J.T. Kiehl, and P. Rosenfeld. 2001. *Science* 294: 2119–2124.

Ramanna, M.V., D.G. Thomas, and S. Varghese. 2001. Estimating nuclear waste production in India. *Current Science* 81: 1458–1462.

Ramakrishnan, P.S., and O.P. Tokay. 1978. Preliminary observation on the impact of *jhum* (shifting) cultivation on the forest ecosystems. In *Proc. Nat. Seminar on Resources Development and Environment in the Himalayan Region*. New Delhi: Department of Science & Technology. 343–354.

Ramasamy, S.M. 2002. Time transgenive tectonics and its control over groundwater regions of Tamil Nadu, India. In *Sustainable development of groundwater resources in semi-arid regions*, ed. Thangarajan M, et al. New Delhi: Oxford. 411–415.

Rao, D.P. 2000. Disaster management. *GIS@Development* 4(3): 28–35.

Rao, D.G., J.C. Katyal, S.K. Sinha, and V. Srinivas. 1995. Impacts of climate change on sorghum productivity in India: Simulation study. In *Climate change and agriculture: Analysis of potential international aspects*. Madison: Amer. Soc. Agronomy. 325–337.

Rao, K.L. 1975. *India's water wealth*. New Delhi: Orient Longman. 255 pp.

Rao, K.R. 2001. Radioactive wastes: The problem and its management. *Current Science* 81: 1534–1546.

Rao, M.H.R. 2001. Development of mining sector in India. *Minerals & Metals Review* (November): 5–11.

Raval, U. 1999. Some recent geophysical constraints from the Central Gondwanaland. *PINSA* 65A: 209–206.

Rawat, J.S., and M.S. Rawat. 1994. Accelerated erosion and denudation in the Nanakosi watershed, Central Himalaya. *Mountain Research and Development* 14: 25–38.

Roy, Arundhati. 1999. *The greater common good*. Mumbai: India Book Distributors. 76 pp.

Roy, Pinaki. 2003. Groundwater level plugging in Bangladesh. *Down to Earth* (April 15) 17–18.

Roy, S.C. 1996. Geoenvironmental studies of the Dehradun–Mussoorie mining belt. In *Contributions in environmental geology*. Kolkata: Geological Survey India. (Pub. No. 43): 80–83.

Roy, S., R.N. Harris, R.U.M. Rao, and D.S. Chapman. 2002. Climate change in India inferred from geothermal observations. *Jour. Geophysical Res*. 107 (B6).

Sah, B.P., K. Honda, and S. Murai. 1979. Land degradation and socioeconomic status: Modelling using remote sensing and GIS for watershed management. *Asia-Pacific Remote Sensing & GIS Journal* 9: 51–61.

Sah, M.P., N.S. Virdi, and S.K. Bartarya. 1996. Slope failure and damming of river channels: Some examples from the Satluj Valley H.P. *Himalayan Geology* 17: 183–191.

Saha, A.K. 1984. *Report on the cyclone review committee* (Vol. 2). New Delhi: Department of Science & Technology. 480 pp.

———, R.P. Gupta, and M.K. Arora. 2002. GIS-based landslide hazard zonation in the Bhagirathi (Ganga) Valley, Himalaya. *Intern. Jour. Remote Sensing* 23: 357–369.

Sahai, A.K. 1999. A brief review of climate change: Causes observations and modelling studies. *Proc. National Acad. Sci. India* LXIX (I): 1–17.

Sahu, K.C. 1995. Arsenic poisoning in Gangetic delta: An anthropogenic model. *Environmental Geology* 25: 1–6.

———. 1998. Impact of residual soil on heavy metal leachate from an ash pond at Korba—an experimental approach. *Ind. Jour. Environmental Protection* 18: 498–504.

Sastry, G., G.P. Juyal, and J.S. Samra. 1995. Conservation measures for sustainable development of degraded lands in Himalaya with special reference to Doon Valley. *Himalayan Geology* 6(2): 47–54.

Selinus, O. 2002. Medical geology: Method theory and practice. In *Geoenvironmental mapping*, ed. P.T. Bobrowsky. Lisse: A.A. Balkema. 473–496.

Selvaraj, K. 2000. Importance of environmental science: Medical geology. *Current Science* 79(1): 8–9.

Shanker, Ravi. 1988. Heat flow map of India and discussion on its geological and economic significance. *Indian Minerals* 42: 89–110.

Sharma, M.L., ed. 1989. *Groundwater recharge*. Rotterdam: A.A. Balkema. 321 pp.

Sharma, S.S. 2004. Mitigating avalanche disasters in Indian Himalaya. In *Coping with natural hazards: Indian context*, ed. K.S. Valdiya. Hyderabad: Orient Longman.

Sharma, V.K. 2000. *India: IDNDR and Beyond*, NCDM. New Delhi: Indian Institute of Public Administration. 94 pp.

———, ed. 2001. *Disaster management*. New Delhi, Indian Institute of Public Administration. 325 pp.

Shekdar, A.V. 1999. Municipal solid waste management—the Indian perspective. *Jour. Ind. Assoc. Environ. Management* 26: 100–199.

Shome, S.K., ed. 1983. *Geotechnical features of major dams in India*. Kolkata: Indian Society of Engineering Geologists. 114 pp.

Sikka, A.K. 2002. Augmentation and Management of groundwater resources: Participatory watershed management as a policy and strategy. In *Sustainable development and management of groundwater resources in semi-arid regions,* ed. Thangarajan, et al. New Delhi: Oxford. 563–969.

Sikka, D.R. 2004. Monsoon drought—a natural hazard in India. In *Coping with natural hazards: Indian context*, ed. K.S. Valdiya. Hyderabad: Orient Longman.

Singh, Gurbachan. 1994. Agroforestry in salt-affected soils In *Proc. Eighth ISCO Conference*. New Delhi. 1466–1483.

Singh, K.P. 2000. Development of waterlogging and salinity in southwestern parts of Punjab state, India. In *Contributions to environmental geoscience*, ed. A.M. Pathan and S.S. Thigale. New Delhi: Aravali Books. 97–106.

Singh, Neelam. 2002. Fig leaf for the business world. *Down to Earth* (March 15): 35–40.

Singh, O.P., T.M.A. Khan, and M.S. Rahman. 2001. Has the frequency of intense tropical cyclones increased in the north Indian Ocean? *Current Science* 80: 575–580.

Singh, Pratap, S.K. Jain, and N. Kumar. 1997. Estimation of snow and glacier melt contribiton to the Chenab river, Western Himalaya, *Mountain Research & Development* 17: 49–56.

Singh, Samar. 1982. *Soil conservation problem, approach and progress in India.* New Delhi: Soil & Water Conservation Division, Ministry of Agriculture. 40 pp.

Singh, Satyajit. 1993. Political economy of large dams: Some tentative assertions. In *Environmental problems and prospects in India*, ed. M. Balakrishnan. New Delhi: Oxford. 195–232.

Singh, Tarlochan. 1995. Artificial recharge of groundwater in Panjab, India. In *Artificial recharge of groundwater*, ed. A.I. Johnson and D.G. Payne. New York: Amer. Soc. Civil Engineers. 831–838.

Singh, T.N. 1992. *Underground winning of coal*. New Delhi: Oxford. 682 pp.

Singh, T.P., and R. Kumar. 1995. *Bhukamp avarodhak ghar nirman*. Dehradun: Lok Vigyan Sansthan. 54 pp.

Singh, V., and D.R. Ghosh. 1975. A plan for development of irrigation in drought-prone areas in India. In *Water for human needs*. New Delhi: Ind. Com. For Water Resources. Vol. 4: 185–195.

Sinha, Rajiv, and K. Jain. 1998. Flood hazards in north Bihar rivers, IndoGangetic Plains. *Memoir Geological Society of India*. 27–52.

Sinha, P.C. 1998. *Desertification.* New Delhi: Anmol Publications. 316 pp.

Sinha, R.C., ed. 1998. *Hydrological disasters.* New Delhi: Anmol Publications. 222 pp.

Soni, V., D.C. Sharma, K.S. Srivastava, and M.S. Sisodia. 1999. Hydrological geophysical and isotope study to trace the course of the buried Saraswati river in Jaisalmer district, Rajasthan. In *Geological evolution of northwestern India*, ed. B.S. Paliwal. Jodhpur: Scientific Publishers. 305–311.

Srinivasan, J., and Sulochana Gadgil. 2002. Asian Brown Cloud—facts and fantasy. *Current Science* 83: 586–592.

Srinivasan, R., and B.L. Sreenivas. 1977. Some new geological features from the Landsat imagery of Karnataka. *Jour. Geol. Soc. Ind*. 18: 589–597.

———, R.U.M. Rao, G.V. Rao., and 5 others. 2001. Radioactivity of Hyderabad granites. Hyderabad: National Geophysical Research Institute. 25 pp.

Srivastava, G.S. 1997. Land degradation due to ravine formation. In *Contributions in environmental geology*. Kolkata: Geol. Survey of India. 189–192.

Srivastava, H.N., and Malti Goel. 1998. Synthesis of meteorological observations and modelling studies to assess climate change mitigation strategies in India. *Current Science* 75: 95–97.

Srivastava, K.N., G.N. Jailte, and J.L. Waza. 1972. Geological factors contributing to the location and design of Pandoh dam, Beas-Satluj link project, Himachal Pradesh. In *Proc. Symp. Geol. & Engineering Problems of River Valley Projects*. Kolkata: Indian Society of Engineering Geologists. 209–220.

Stille, H., J. Brantmark, A. Heiner, and J. Martna. 1998. Evaluation of squeezing experiences in the Uri Project India. In *Hydropower development in the Himalayas*, ed. V.D. Choubey. New Delhi: Oxford. 281–304.

Subramanian, V. 2000. *Water: Quantity-Quality perspective in South Asia.* Surrey: Kingston Intern. Publishers. 256 pp.

———. 2001. Water: Quantity and Quality. In *Ecohydrology*, ed. V. Subramanian and A.L. Ramanathan. New Delhi: Capital Publishing Company. 241–264.

———, and R.A. Dalavi. 1978. Some aspects of stream erosion in the Himalaya. *Himalayan Geology* 8: 822–834.

———, and A.L. Ramanathan. 1996. Nature of sediment load in the Ganges-Brahmaputra river system in India. In *Sea-level Rise and Coastal Subsidence*, ed. J.D. Milliman and B.U. Haq. Amsterdam: Kluwer Acad. Pub. 151–168.

Sukhija, B.S., D.V. Reddy and P. Nagabhushanam. 1998. Isotopic fingerprints of palaeoclimates during the last 30000 years in deep confined groundwaters of Southern India. *Quat. Res*. 50: 252–260.

Suma Latha, S., S.R. Ambika and S.J. Prasad. 1999. Fluoride contamination status of groundwater in Karnataka. *Current Science* 76: 730–734.

Summers, I. 1998. Groundwater recharge: An overview of estimation problems and recent developments. In *Groundwater pollution aquifer recharge and vulnerability*, ed. N.S. Robins. London: Geological Society. 107–115.

Sundaraiya, E., and V.N. Wakpanjar. 1996. Reservoir created in karstic environment. In *Proc. Symp. Earthen Dams*, ed. C.V.J. Varma and A.R.G. Rao. New Delhi: Central Board of Irrigation & Power. 349–362.

Suresh, D.S., M. Babu, V. Nandakumar, B. John, B.K. Jayaprasad, and S.V. Pramod. 2000. Siltation analysis in Neyyar reservoir and forest degradation in its catchments: A study from Kerala State, India. *Environ. Geology* 39: 391–397.

Susheela, A.K. 1999. Fluorosis management programme in India. *Current Science* 77: 1250–1256.

Talwani, Pradeep. 2000. Seismogenic properties of the crust inferred from recent studies of reservoir-induced seismicity—Application to Koyna. *Current Science* 79: 327–1333.

———, and Kusala Rajendran. 1991. Some seismological and geometric features of the intraplate earthquakes. *Tectonophysics* 186: 19–41.

Tangri, A.K. 2000. *Integration of remote sensing data with conventional methodologies in snow-melt runoff modelling in Bhagirathi river basin, U.P. Himalaya* (unpublished report). Lucknow: Remote Sensing Application Centre. 601.

Tejwani, K.G., R. Dayal, U. Verma, H.M. Godwani, and S. Gupta. 1975. Classification control and reclamation of ravine lands. In *Soil and water conservation research* , ed. K.G. Tejwani, et al. New Delhi: Indian Council of Agricultural Research. 254–310.

Tett, S.F.B., P.A. Stott, M.R. Allen, and 2 others. 1999. Causes of twentieth-century temperature change near earth's surface. *Nature* 399: 569–571.

Thampi, P.K. 1997. *Evaluation study in terms of landslide mitigation in parts of Western Ghats, Kerala*. Trivandrum: Centre for Earth Science Studies. 33 pp.

Thangarajan, M., S.N. Rai, and V.S. Singh. 2002. *Sustainable development and management of groundwater resources in semiarid regions with special reference to hard rocks*. New Delhi: Oxford. 587 pp.

Thapliyal, V., and S.M. Kulshrestha. 1993. Recent climate changes and trends over India. *Tisglow* 4(1): 7–17

Thigale, S.S., and A. Khandge. 2000. Landslides in Western Ghats of Maharashtra. In *Contributions to environmental geoscience*, ed. A.M. Pathan and S.S. Thigale. New Delhi: Aravali Books Intern. 161–174.

Thiruvengadachari, S., S. Jonna, K.A. Hakeem, and 6 others. 1996. Improved water management: The IRS-1C. *Current Science* 70: 589–599.

Tolia, D.S. 1997. Preventive and control of mass-wasting in the Himalayan region, particularly along highways. In *Perspectives of mountain risk engineering in Himalayan region*, ed. D.K. Agrawal, A.P. Krishna, V. Joshi, K. Kumar, and L.M.S. Palni. Nainital: Gyanodaya Prakashan. 143–161.

Uprety, B.K. 1988. Impact of riverbed erosion flood hazard and river-shifting in Nepal. In *Proc. Intern. Symp. Impact of Riverbed Erosion Flood Hazard and Problems of Population Displacement*, Dhaka. 27 pp.

Valdiya, K.S. 1985. Accelerated erosion and landslide-prone zones in the Himalayan region. In *Environmental regeneration in Himalaya: Strategy and concept*, ed. J.S. Singh. Nainital: Gyanodaya Prakashan. 12–38.

———. 1987. *Environmental geology: Indian context*. New Delhi: Tata-McGraw Hill. 583 pp.

———. 1992. Must we have high dams in the geodynamically active Himalayan domain? *Current Science* 62: 289–296.

———. 1997a. High dams in central Himalaya in context of active faults, seismicity and societal problems. *Jour. Geol. Soc. Ind*. 49: 479–494.

———. 1997b. *Developing a paradise in peril*. Kosi-Katarmal: 17th G.B. Pant Memorial Lecture, G.B. Pant Institute of Himalayan Environment & Development. 26 pp.

———. 1998. Late Quaternary movements and landscape rejuvenation in southeastern Karnataka and adjoining Tamil Nadu. *Jour. Geol. Soc. Ind*. 51: 139–166.

———. 1999. Why does Brahmaputra remain untamed? *Current Science* 76: 1301–1305.

———. 2001. Tectonic resurgence of the Mysore Plateau and surrounding regions in cratonic southern India. *Current Science* 81: 1068–1089.

———. 2002. *Saraswati: The river that disappeared*. Hyderabad: Universities Press. 116 pp.

———. 2004. Geodynamic perspective of natural hazards in India. In *Coping with natural hazards: Indian context*, ed. K.S. Valdiya. Hyderabad: Orient Longman.

Valiathan, M.S. 1998. *Soil in heart*. D.N. Wadia Honour Lecture. Dehradun: Wadia Institute of Himalayan Geology. 10 pp.

Varma, O.P. 1999. *Emerging trends in mineral exploration and valuation*. Roorkee: Indian Geological Congress. 21 pp.

Vaughan, D.G. 2001. Devil in the detail. *Science* 293: 1777–1779.

Vohra, C.P. 1978. Glacier resources of the Himalaya and their importance to environmental studies. In *Proc. Nat. Seminar Resources Development and Environment in the Himalayan Region*. New Delhi: Department of Science & Technology. 551–459.

———. 2000. Glaciology, 25 years ago. In *Research highlights in earthscience*, ed. O.P. Varma and T.M. Mahadevan. Roorkee: Indian Geological Congress. 199–206.

Vyas, J.N., and R. Parthasarathy. 2001. *Saket environmental handbook*. Ahmedabad: Saket. 455 pp.

Warrick, R.A., and Q.K. Ahmad, eds. 1997. *The implication of climate and sea level changes for Bangladesh*. Dordrecht: Kluwer Academic Publishers. 415 pp.

Watson, I., and A.D. Burnett. 1995. *Hydrology: An environmental approach*. Cambridge: Buchanan Books. 4–29.

Watson, R.T., M.C. Zinyowera, and R.H. Moss. 1998. *The regional impact of climate change*. Cambridge: Cambridge University Press. 514 pp.

Yeats, R.S., and V.C. Thakur. 1998. Reassessment of earthquake hazard based on a fault-bend-fold model of the Himalayan boundary fault. *Current Science* 74: 230–233.

Young, Anthony. 1998. *Land resources: Now and for the future*. Cambridge: Cambridge University Press.

Details for the following references were unavailable while going to press. They will be added in future editions:

Anwar et al, 2002; Braun, 1991; Gupta, 1998; Nanda et al, 1998; Sharma, 1997.

# Index

**Related Books by Orient Longman and Universities Press**

*Climate Change and India: Vulnerability Assessment and Adaptation*
P.R. Shukla, Subodh Sharma, N.H. Ravindranath, Amit Garg and Sumana Bhattacharya

*Climate Policy Assessment for India: Applications of Asia–Pacific Integrated Model (AIM)*
P.R. Shukla, Ashish Rana, Amit Garg, Manmohan Kapshe and Rajesh Nair

*Coping with Natural Hazards: Indian Context*
K.S. Valdiya (ed.)

*Crowded Greenhouse*
John Firor and Judith Jacobsen

*Disaster Management*
Harsh K. Gupta (ed.)

*The Earth Policy Reader*
Lester P. Brown, Janet Larsen and Bernie Fischlowitz-Roberts

*Eco-Economy: Building an Economy for the Earth*
Lester P. Brown

*Eco-Socialism or Eco-Capitalism?: A Critical Analysis of Humanity's Fundamental Choices*
S. Sarkar

*Environmental Consciousness and Urban Planning*
M.N. Buch

*Global Warming: Can Civilization Survive?*
P. Brown

*Greenhouse: The 200-year Story of Global Warming*
G.E. Christianson

*Joint Forest Management in India*
N.H. Ravindranath and P. Sudha

*The Naturalisation of Scarcity: The Politics and the Poetics of Water in Western India*
Lyla Mehta

*Regional Land Cover Changes, Sustainable Agriculture and their Interaction with Global Change*
V. Ravichandran (ed.)

*Rio, Johannesburg and Beyond: India's Progress in Sustainable Development*
LEAD India

*Universities Press Dictionary of Earth Science*
John O.E. Clark and Stella Stiegeler

*Water for Pabolee: Stories about People and Development in the Himalaya*
Robert C. Alter

Wageningen University Water Resources Series

*Democratising Micro-Hydel: Structures, Systems and Agents in Adaptive Technology in the Hills of Nepal*
Amreeta Regmi

*Engineering Participation: The Process and Outcomes of Irrigation Management Transfer in the Terai of Nepal*
Pushpa Raj Khanal

*Institutions, Technology and Water Control: Water Users Associations and Irrigation Management in Two Large-scale Systems in India*
Vishal Narain

*On the Waterfront: Water Distribution, Technology and Agrarian Change in a South Indian Canal Irrigation System*
Peter P. Mollinga

*Social Designs: Tank Irrigation Technology and Agrarian Tranformation in Karnataka, South India*
Esha Shah

*Water, Works and Wages: The Everyday Politics of Irrigation Management Reform in the Philippines*
Joost Oorthuizen

Books by K.S. Valdiya published by Universities Press

*Coping with Natural Hazards: Indian Context*

*Dynamic Himalaya*

*Himalaya: Emergence and Evolution*

*Saraswati: The River that Disappeared*